Radio Utopia

THE HISTORY OF COMMUNICATION

Robert W. McChesney
and John C. Nerone, editors

*A list of books in the series
appears at the end of this book.*

Radio Utopia

Postwar Audio Documentary in the Public Interest

MATTHEW C. EHRLICH

UNIVERSITY OF ILLINOIS PRESS
Urbana, Chicago, and Springfield

© 2011 by the Board of Trustees
of the University of Illinois
All rights reserved
Manufactured in the United States of America
C 5 4 3 2 1
♾ This book is printed on acid-free paper.

Library of Congress Cataloging-in-Publication Data
Ehrlich, Matthew C.
Radio utopia : postwar audio documentary in the public interest /
Matthew C. Ehrlich.
  p. cm. — (The history of communication)
Includes bibliographical references and index.
ISBN-13: 978-0-252-03611-8 (hardcover : alk. paper)
ISBN-10: 0-252-03611-5 (hardcover : alk. paper)
1. Radio broadcasting—United States—History—20th century.
2. Documentary radio programs—United States—History—
20th century. 3. Radio broadcasting—Social aspects—United
States—20th century. 4. Radio broadcasting—Political aspects—
United States—20th century. I. Title.
PN1991.3.U6E37   2011
791.44—dc22   2010040243

Dedicated to George Ehrlich (1925–2009)
Father, teacher, scholar

To despair of the world is to resign from it.

—Norman Corwin

Contents

Acknowledgments ix

Introduction: Utopian Dreams 1
1. A Higher Destiny 13
2. One World 24
3. New and Sparkling Ideas 46
4. Home Is What You Make It 71
5. The Quick and the Dead 104
6. Hear It Now 129
7. Lose No Hope 155

Notes 165
Index 211

Acknowledgments

Thanks to the staffs of the following archives, who made researching this book such a pleasure: CBS News Archives Reference Library, New York; Daley Library Special Collections, University of Illinois at Chicago; Dolph Briscoe Center for American History, University of Texas at Austin; Howard Gotlieb Archival Research Center, Boston University; Library of American Broadcasting, University of Maryland at College Park; Library of Congress, Washington, D.C.; Paley Center for Media, New York and Beverly Hills; Rare Book and Manuscript Library, Columbia University; Ratner Center for the Study of Conservative Judaism, Jewish Theological Seminary, New York; Robert F. Wagner Labor Archives, Tamiment Library, New York University; Special Collections, Thousand Oaks Library, Thousand Oaks, California; University of Illinois Archives and University of Illinois Library, Urbana; Wesleyan Cinema Archives, Wesleyan University, Middletown, Connecticut; and the Wisconsin Historical Society, Madison.

Equally pleasurable, if not more so, was the privilege to speak with a few of those who lived the actual events that this book relates. Thanks to Norman Corwin, Robert Lewis Shayon and his wife Nash Cox, and Ruth Ashton Taylor.

For access to historic radio broadcasts, thanks to Digital Deli Online (http://www.digitaldeliftp.com); Jerry Haendiges Productions, Whittier, California (http://www.otrsite.com); Original Old Radio, Berea, Kentucky (http://www.originaloldradio.com); and the Radio Program Archive, University of Memphis, Tennessee (https://umdrive.memphis.edu/mbensman/public).

For helpful comments concerning earlier drafts of this book, thanks to Victor Pickard, Patrick Washburn, and Mike Conway. Thanks as well to Kendra Boileau and Daniel Nasset of the University of Illinois Press and to John Nerone and Robert McChesney of the History of Communication series.

For tips, shelter, and good cheer, thanks to Joe Saltzman, Chris Chandler, and Kavitha Cardoza. For continuing to provide an intellectually vital place to investigate and think, thanks to colleagues and students at the College of Media at the University of Illinois at Urbana-Champaign. Finally, thanks and love always to my family.

Radio Utopia

INTRODUCTION

Utopian Dreams

It was the spring of 1945, so the story goes, and Edward R. Murrow was holding court among a group of his colleagues in war-ravaged Europe. During World War II, radio journalism had come into its own. Murrow had become internationally renowned during the German Blitz against London prior to America's entry into the war. According to the poet Archibald MacLeish, Murrow's CBS radio dispatches had demolished in Americans' minds "'the ignorant superstition that violence and lies and murder on another continent are not violence and lies and murder here.'"[1] Murrow had continued to inform his fellow citizens about Nazi brutality, most recently via a graphic radio report about the Buchenwald concentration camp: "If I've offended you by this rather mild account of Buchenwald, I'm not in the least sorry."[2] During the same period, he had helped assemble a celebrated group of reporters for CBS, the so-called Murrow Boys, most of whom were indeed men—Eric Sevareid, William Shirer, Charles Collingwood, Howard K. Smith, and others. Some from among that group were in the room with Murrow now, all of whom "had made the antifascist cause their own, buoyed by a sense of unity at home," as one of Murrow's biographers later put it.[3]

Also in the room was Robert Lewis Shayon, one of several radio writers and directors who had helped bolster that sense of unity. Shayon said that the Depression had made him and others "sensitive and sympathetic to justice, social 'causes,' and reform."[4] That sensitivity had carried over to the war, during which he and his peers produced programs vilifying the enemy abroad while warning against injustice at home. Norman Corwin had helped lead the way by airing installments of the series *An American in England* live via shortwave from London. Shayon had come to Europe as part of a War Department–sponsored tour giving other radio dramatists firsthand knowledge of how the battle was progressing. Among those accompanying him was William Robson, the author of the CBS program *Open Letter on Race Hatred*, which had blisteringly criticized the conditions that triggered a deadly wartime riot in Detroit.[5] Now the war was ending, and Shayon listened as Murrow extolled his assembled colleagues in Europe to carry on the good fight back home: "'We've seen what radio can do for the nation in war. Now let's go back to show what we can do in *peace!*'"[6]

This book is the story of what happened next. Journalists joined dramatists in using radio to try to remake America and the world for the better. Murrow helped form the CBS Documentary Unit with Shayon as a member, and similar efforts developed at the other networks. They produced programs advocating action on everything from juvenile delinquency, slums, and race relations to venereal disease, atomic energy, and arms control. For a time, their efforts were encouraged by the commercial broadcasting industry, which was under pressure from the Federal Communications Commission (FCC) to demonstrate that it was truly serving the public interest. The head of the CBS Documentary Unit, Robert Heller, hailed the emergence of "a virtual Utopia for craftsmen who believe in radio's usefulness as a social force."[7] By 1951, that "utopia" had evaporated as radio gave way to television, the war against fascism gave way to the cold war against communism, and many of radio's most acclaimed "craftsmen"—including Heller, Shayon, Corwin, and Robson—landed in the pages of the red-baiting publication *Red Channels*, their careers never to be the same again.

Interpretive Framework

The media landscape underwent an extraordinary transformation between 1945 and 1951. As one account has put it, "[A] small radio system dominated by four networks" was replaced by "a far larger AM-FM radio and television system in which networks concentrated on television and left radio stations to their own programming resources."[8] Ambitious radio programming endeavors launched immediately after the war were largely abandoned six years later. Edward R. Murrow, the great champion of radio and skeptic of television, moved to the new medium as of November 1951 with *See It Now*, marking the end of an era.

In the interim, American audio documentary would enjoy a brief heyday that vividly reflected the social and cultural climate of the times. That heyday has yet to receive the attention it deserves. Writing in 1965, A. William Bluem acknowledged radio's achievements in the six years immediately following the war: "From experience gained in earlier experiments it had evolved an authentic and dramatic form of journalistic documentary, dealing with the crises of the world as they continued to arise. It had worked from dramatic restatement of fact to drama made *with* fact. It had presented information in a compelling form on numberless major and minor issues and problems confronting the American people.... And as it did all these things, it gave a legacy to television."[9]

Bluem pinpointed several reasons why postwar audio documentary is significant. It developed though early reality-based radio shows such as *The March of Time* and the socially conscious programs of the war years. It then underwent a crucial transformation as docudramas featuring actors impersonating real people were supplanted by actuality-based programs that took advantage of new recording technology. Along the way, it showcased the talents of noted journalists and writers grappling with some of the era's thorniest concerns. Norman Corwin produced the series *One World Flight*, which took him on a round-the-world trip assessing the prospects for postwar peace. Robert Lewis Shayon created *The Eagle's Brood* on delinquent youth as well as the historical series *You Are There*. Erik Barnouw—later a pioneering historian of broadcasting and documentary—wrote *V.D.: The Conspiracy of Silence* about syphilis. Ruth Ashton talked to Albert Einstein in researching *The Sunny Side of the Atom* on atomic energy. Fred Friendly tackled the same subject with *The Quick and the Dead*, starring Bob Hope. Friendly then joined Murrow in creating *Hear It Now* for CBS, which evolved into the venerated television series *See It Now*.

As such efforts shifted to television, American radio was left as "a shell of a medium" in terms of documentary, according to Bluem.[10] Consequently, Bluem focused the brunt of his 1965 study on TV, as have others in subsequent years. A growing scholarly literature has called attention to the network-television documentary as the "product of converging social, economic, political, institutional, and discursive forces."[11] For example, one study shows how a series of documentaries in the early 1960s emerged from a consensus among government officials and broadcasters that television ought to raise its standards and promote the image of the United States as a bastion of freedom during the cold war. Another study analyzes how muckraking documentaries of the 1960s and 1970s were enabled by government regulation that encouraged such programs even as they sparked government investigations into their reporting methods.[12]

The historical evolution of American radio documentary has received much less scrutiny. One broadcast-journalism historian briefly discusses a few key works such as *The Eagles' Brood*, whereas another notes that some of those programs "are still remembered as the pinnacle of radio writing and production."[13] Even so, Christopher Sterling and Michael Keith argue that historians have overlooked much of the serious radio work produced prior to television's ascendancy.[14]

At the same time, however, the radio documentary has experienced a resurgence in recent years. A new wave of work beginning in the 1990s has attracted notice for "telling stories that can't be told on film" and introducing listeners to those with "lives that are different than theirs."[15] Relatively inexpensive recording technology and editing software have made radio documentary an increasingly democratic and accessible medium. One study describes contemporary audio documentary as a form of ethnography that promotes civic life. The study's authors connect that work to earlier documentaries on radio and on records such as those produced by John and Alan Lomax for the Library of Congress before the war and by Tony Schwartz for Folkways Records after the war.[16]

Audio documentary's revitalization has paralleled a rise in radio-history studies.[17] The historians Michele Hilmes and Susan Douglas have each analyzed the heyday of American radio in relation to Benedict Anderson's concept of "imagined communities" of citizens separated by geography but bound together by mass media, particularly newspapers. Douglas argues that radio played a greater role than print in forming such communities, even as it also allowed listeners to experience "multiple identities," some of which were "completely allied with the country's prevailing cultural and political ideologies" and others of which were "suspicious of or at odds with official culture."[18] In a similar vein, Susan Merrill Squier notes radio scholars' particular interest in the medium's dual nature—"its power to enforce the status quo (especially consumerism and stereotyping of race, gender, and ethnicity)" alongside "its capacity to provide a voice for resistance and critique."[19]

This book builds upon those related strands of research at the intersection of radio studies, documentary studies, and journalism studies. The postwar programs pointed to tensions and contradictions intrinsic to both radio and documentary. As Bluem notes, they showed "that radio had finally reached its goal of making documentary a force to influence a vast listening audience" by exposing social ills;[20] at the same time, they stopped short of calling for radical change. They demonstrated how commercial and regulatory pressures both facilitate and constrain documentary, as would be the case with the network-television documentaries that followed in their wake. They also showed what has been described as documentary's enduring conflict "between the claim to truthfulness and the need to select and represent the reality one wants to share."[21] Technological limitations and a network recording ban compelled many radio programs of the period to employ dramatizations, and hence they were akin to what today would be labeled "docudrama." But they were called "documentaries" in their time, and their claims to truth are no different from those of the documentaries rooted in recorded actuality that had become the norm by 1951. (For example, the series *CBS Is There*—later and better known as *You Are There* and noted for its dramatizations of history—was produced by the CBS Documentary Unit near the end of its radio run.)

The programs revealed tensions within journalism as well, as highlighted in contrasting essays by James Carey and David Paul Nord. In an often cited 1974 piece, Carey declared that historians should move beyond "Whig"-like accounts of journalism's inexorable progress and improvement and instead try to reveal "historical consciousness" by showing how it felt "to live and act in a particular period."[22] Carey also wrote that rather than reducing all consciousness to the level of ideology or viewing journalism as the transmission of information to a passive public, scholars should see journalism "as an exercise in poetry and utopian politics."[23] In response, Nord argued that historians ought not to get so caught up in utopian views as to overlook that "the 'consciousness' embedded in the language of journalism is the product of large institutions" and "the exercise of power." In brief, press history should retain a "focus on powerful individuals and institutions."[24]

Both perspectives illuminate postwar audio documentary. The programs expressed a consciousness consistent with the tradition of the American "reformist Left" that the philosopher Richard Rorty said was active for much of the twentieth century and that encompassed "lots of people who called themselves 'communists' and 'socialists,' and lots of people who never dreamed of calling themselves either." What they shared were "utopian dreams—dreams of an ideally decent and civilized society."[25] Those dreams animated radio writers in the "Cultural Front" that advanced progressive causes during the 1930s and 1940s.[26] As one historian describes it, they shared a vision of "a world without prejudice, brutality, political oppression, or economic deprivation" and sought to produce programs that "challenged Americans to reform their own country."[27] Those programs expressed their own form of poetry in the service of utopian politics.

At the same time, such programs existed only because they temporarily aligned with the interests of powerful individuals and institutions. The commercial networks used the programs to help deflect criticism from government regulators and to distinguish themselves from their competition. The postwar radio documentaries were in part a response to the FCC's 1946 "Blue Book" that criticized radio's rampant commercialism and outlined public-service standards for broadcast licensees.[28] The documentaries aired at irregular and less-than-ideal times, and once television and McCarthyism took hold, they went into eclipse altogether. All that is consistent with Nord's view of history.

Indeed, one could broadly interpret the story of postwar audio documentary in a couple of ways, each resting on the notion that power trumps utopianism. The first interpretation would be of a utopia that was lost, a golden age of invention, independence, and hope done in by greed and reactionaryism. Norman Corwin recalled seeing friends "exiled, punished, jailed, ostracized" by the blacklist while witnessing the decline of his chosen medium of radio: "I had been riding a wonderful charger—a beautiful horse, the saddle and equipage of which was furnished by a great network [CBS]—and that horse was shot out from under me."[29] Robert Lewis Shayon, who became a TV critic after leaving CBS, wrote upon the 1965 death of Edward R. Murrow that his "passing symbolized the end of a great adventure in broadcasting," even though "the broadcasting idealism that Murrow represented died many years ago."[30] The death of that idealism is a major theme of Murrow's biographers, with one ironically contrasting his death in exile from journalism with his exuberant exhortation to his colleagues in Europe twenty years earlier: "Now let's go home to show what we can do in *peace!*"[31] Such accounts underscore Murrow's status as the "patron saint" of broadcast news and as a tragic, martyr-like figure.[32]

A second broad interpretation of postwar documentary would be of a utopia that never was. Such an interpretation is less likely to see Murrow as a saint than as an overrated "glory hog who played it safe, more puffery than paladin," and as a figure whose historical importance has been greatly exaggerated.[33] (Even Murrow's purported "let's show what we can do!" declaration is questionable; Shayon's diary of his 1945 European trip recounts meeting with Murrow but does not mention such a dramatic pronouncement.[34]) Similarly, the radio works of the Cultural Front can be viewed as not having aged well and as today sounding "trite," "overwrought," "awkwardly bombastic," and filled with "unwitting condescension."[35] Although they condemned social prejudice, they "still kept the reins of communication in the hands of the white majority" that also largely excluded women.[36] The earnest public-affairs news programs to which they gave rise privileged journalistic authority as opposed to serving what has been called the "democratic impulse at the heart of documentary, to let people speak for themselves."[37] As for the decline of radio documentary and the rise of the blacklist, both are unsurprising given broadcasting's cooptation by corporate America prior to the war and its use as a propaganda instrument during the war itself, whereas the postwar years saw the media harnessed in favor of a massive surge in consumerism at the same time that liberal-minded efforts at media reform were squelched.[38]

Each of those two interpretations is valid in its way. As this book will relate, postwar audio documentary did serve corporate concerns, at least for a time. That time would pass, and there would be tragic or at least disillusioning consequences for some of the participants. The book's title, *Radio Utopia*, thus can be read ironically. Nevertheless, the title also can be taken in earnest, for utopian sensibilities did underlie the era's documentary. In his memoirs many years later, Shayon commented that contemporary observers "who view the world with a more cynical realpolitik attitude" might find such an outlook hopelessly romantic. Still, it was genuine: "idealism in the flush of military triumph over evil—amid the sense that a new world was about to be born."[39] Whatever the shortcomings or limitations of the era's programs and the people who created them, they were rarely cynical. Even if the anecdote about Murrow urging his peers to show Americans what radio could accomplish in the name of peace did not happen exactly the way it was described, the record suggests that in fact the radio journalists and dramatists did do their best to "show them." That also is a key part of this book's story.

Plan of Procedure

The book focuses almost entirely on long-form radio storytelling that appeared in single programs or as part of a series; daily newscasts and stand-alone commentary are not included. The term "audio documentary" is used rather than "radio documentary" to place the era's programs within a long, ongoing tradition of such work that has appeared not only on radio but also on records and the Internet. One work discussed briefly—Fred Friendly and Edward R. Murrow's *I Can Hear It Now*—originally was released on records.

"Documentary" here is viewed broadly to include the wide range of programs that were labeled as such at the time. That is in keeping with an understanding of documentary that goes well beyond the stereotype of the public-affairs program centering around expert perspectives and stentorian narration.[40] William Stott has described documentary as being devoted to "the presentation of actual facts in a way that makes them credible and telling to people," as well as to the belief "that the world can be improved and yet must be celebrated as it is."[41] A wide range of programs met that definition immediately following the war, including those incorporating actors, sound effects, and music. The shift away from such dramatizations paralleled a shift toward an era of "social responsibility" and what has been called "high modernism" in journalism.[42] Regardless, the implicit assumption persisted that documentaries were intended to serve the public interest, as the federal government mandated broadcasters to do. "Public interest" always has been an amorphous concept, but the FCC commissioner Clifford Durr's understanding of it as of 1947 remains relevant today: "[S]erving the ends of a democratic society" by facilitating "an informed and understanding people capable of making choices."[43] Significantly, the documentaries also served the commercial networks' interest by burnishing their corporate image in the face of government scrutiny and public criticism.

"Postwar" in this context again refers to the years 1945–51, although radio documentaries obviously continued to be produced after then. The book does not attempt to be comprehensive. It examines American network documentary as opposed to programs produced by local stations or other countries, and it concentrates on selected key programs and figures, inevitably leaving out others. (For example, the focus is primarily on CBS, ABC, and NBC, although Mutual did many documentaries of its own.)

Chapter 1 examines the context from which postwar audio documentary emerged. Forerunners included 1930s series such as *The March of Time* and *Cavalcade of America*, which were created in large part either as corporate advertisements (with *March of Time* promoting *Time* magazine) or as public relations (with *Cavalcade of America* enhancing DuPont's image in the face of charges that it had acted as a "merchant of death" during World War I). Other programs were "sustaining" or nonsponsored. Although the networks created them partly as a sop to federal regulators, they granted airtime to an idealistic cadre of writers including Norman Corwin, who believed that radio should serve a higher purpose than merely peddling commercial goods.

Chapter 2 looks more closely at Corwin's work. He produced a number of shows attacking fascism prior to and during World War II, culminating with his classic *On a Note of Triumph*, which aired upon the Nazi surrender in Europe. The following year, he undertook a global tour in the name of Wendell Willkie, the onetime Republican presidential nominee whose book *One World* had been a bestseller. Corwin took along a wire recorder and incorporated the interviews he collected into a thirteen-part CBS series called *One World Flight* that called for global peace as the cold war began to intensify. The scheduling of the series against Bob Hope's popular program generated controversy.

Chapter 3 focuses on the CBS Documentary Unit. William Paley, the chief of CBS, had called for innovations in radio programming in partial response to the 1946 FCC Blue Book report, even as the broadcasting industry vociferously attacked the report. The documentary unit was one result. Its first major production, *The Eagle's Brood*, was written by Robert Lewis Shayon following a nationwide trip that he took to investigate juvenile delinquency. It culminated with a visit with the neighborhood organizer Saul Alinsky in Chicago, and the resulting program advocated neighborhood councils as a solution to delinquency. Another unit production, *The Sunny Side of the Atom*, pondered the potentially positive effects of atomic science. It too rested on several months of research, this time undertaken by Ruth Ashton, one of the few women invited to participate in creating the documentaries. The CBS unit's programs featured actors such as Joseph Cotten and Agnes Moorehead, and the shows became the focus of a network research endeavor incorporating the "motivation research" of Ernest Dichter and the "Program Analyzer," co-invented by Paul Lazarsfeld and the CBS president Frank Stanton.

Chapter 4 looks at documentary efforts at NBC and ABC. The chapter title, "Home Is What You Make It," is borrowed from an early series written by NBC's Lou Hazam and points to the documentaries' calls for American self-improvement. Hazam went on to write *Living 1948*, which examined everything from solar energy to education reform. The program regularly featured George Gallup and his pollsters, at least until Gallup incorrectly predicted the winner of the 1948 presidential race. At ABC, Robert Saudek oversaw the production of programs such as *Slums*, the unique musical documentary *1960?? Jiminy Cricket!*, and *V.D.: The Conspiracy of Silence*, with Erik Barnouw writing the latter show. Saudek also was the impetus behind *Communism—U.S. Brand*, an ominous warning about domestic subversion. The program triggered bitter denunciations from some on the left and led to a memorable exchange of letters between the program's author, Morton Wishengrad, and his fellow radio writer Norman Rosten. Meanwhile, at CBS, Robert Lewis Shayon created the series *CBS Is There*, which purported to place real-life radio news reporters at the scene of re-created historical events. The show's blend of fact and fiction and the participation of the network's journalists discomfited Edward R. Murrow even as he created his own historical documentary with Fred Friendly for Columbia Records, *I Can Hear It Now*.

Chapter 5 examines the dramatic changes in the radio landscape that began in 1948 and accelerated in the years that followed. Norman Corwin departed CBS after William Paley said that the network was facing mounting pressure over its perceived liberalism and that it no longer would support programs that failed to appeal to a wide audience. Robert Lewis Shayon and others also left CBS following budget cutbacks, with television increasingly absorbing network attention and revenues. Following the defeat of the Progressive party candidate Henry Wallace in the 1948 presidential race and the communist takeover of China the following year, pacifism fell increasingly into disfavor, and liberalism shifted accordingly, as advocated by Arthur Schlesinger Jr.'s book *The Vital Center*. Audio documentary followed suit as the Twentieth Century Fund and the onetime Yale chaplain Elmore McKee created the series *The People Act* about Americans' ability to solve their own problems and uphold democratic traditions. Fred Friendly's *The Quick and the Dead* aired on NBC just after the outbreak of the Korean War and looked at America's development of atomic weapons. Friendly's program and *The People Act* to varying degrees employed the new technology of audiotape that was transforming documentary.

Chapter 6 begins by looking at the publication of the red-baiting book *Red Channels*, which listed several leading radio writers and led to their blacklisting. At the same time, Friendly was departing NBC for CBS and a permanent partnership with Murrow. They created the series *Hear It Now*, which made full use of Friendly's production skills with taped actualities, in the process leaving the days of dramatized documentary behind. The series aired during a tumultuous six months that coincided with the grimmest stages of the Korean War, the firing of Douglas MacArthur, the Kefauver committee hearings on organized crime, and the rise of McCarthyism. The series' final installment in June 1951 featured Murrow's angry response to McCarthy, four years before the journalist's well-known television program on the Wisconsin senator. By the end of 1951, Murrow and Friendly had moved to television—Friendly with enthusiasm, Murrow with foreboding. Chapter 7 considers the aftermath and legacy of the postwar audio-documentary era while considering its ramifications for journalism, radio, and documentary.

The book draws upon the collected papers of several principal players, including Murrow, Friendly, Corwin, Shayon, Wishengrad, and others. It also draws upon original audio recordings, scripts, and notes culled from a number of university archives, the CBS News Archives, the NBC company records at the Wisconsin Historical Society, the Paley Center for Media in Beverly Hills and New York City, and various private collectors and online sites. Context is provided by accounts in the day's popular press and trade press, especially *Variety,* which regularly covered the networks and their documentary efforts.

The goal here is to reveal how audio documentaries responded to the political, economic, and cultural upheaval of the era and how they highlighted what James Carey termed the enduring "moral and political ambiguities of journalism."[44] Robert Lewis Shayon hinted at those ambiguities as he reflected upon his career near the end of his life. "I was neither victim nor hero," he wrote in his memoirs. "I was a human being, as all of us were, trapped in the tangle of peculiar times."[45] This book will try to capture the times' peculiarities while avoiding a facile tale of heroes and villains. It will tell of hubris, naiveté, and cowardice alongside aspiration, determination, and hope. We need to understand both sides of the tale if we are to learn its lessons and fruitfully apply them to the present.

1

A Higher Destiny

Utopian hopes for American radio are almost as old as the medium itself. In the words of one historian, radio was widely seen in the early 1920s as "a force new and powerful that seemed to have unlimited possibilities for social good" and as a "bright hope for a better world." It would promote peace and democracy while raising the standards of education and mass culture. However, disillusionment soon followed over "the utter dullness, banality, and absence of controversy" offered by the commercial system that quickly came to dominate American broadcasting.[1] Hoping to reverse the trend, a coalition of labor, educational, and religious groups lobbied Congress to reserve a significant share of the airwaves for nonprofit entities. The broadcasting industry successfully fought to kill the initiative, and by the mid-1930s, the continued corporate control of radio was ensured.[2]

Prewar Documentary and News

Even within that corporate framework—which, according to one radio critic of the day, diminished the medium "'to the level of a gigantic vaudeville show'"[3]— documentary-type programming soon appeared. The earliest examples were commercial ventures. *The March of Time* debuted on CBS in 1931 and was underwritten by *Time* magazine, which developed the program out of experimental news dramatizations called *NewsCasting* and *NewsActing*. *Time*'s new radio series aggressively promoted its parent. "There is one publication which watches, analyzes, and every seven days reports the march of human history on all the fronts," an announcer intoned at the start of the program's premiere. "Tonight the editors of *Time*, the weekly newsmagazine, attempt a new kind of reporting of the news, the reenacting as clearly and dramatically as the medium of radio will permit [of] some themes from the news of the week." There followed dramatizations of "Big Bill" Thompson's victory in the Chicago mayoral primary, the death of the *New York World* newspaper, and the transfer of French prisoners to Devil's Island, among other happenings.[4] A subsequent 1931 episode reenacted an encounter between "India's first citizen—world-famed Mahatma Ghandi" and a starstruck American tourist, Mrs. Hattie Belle Johnston ("Oh, Mahatma, when are you coming to America? They'll go wild about you there"). The episode ended by asking whether "threats to end the life of nut-brown little Mahatma Gandhi" would be realized.[5]

The March of Time lost money for *Time* at first, and the publishing company attempted to cancel it. Listeners protested, other sponsors were eventually found, and the series remained on radio (with brief hiatuses) until 1945. It featured top radio actors such as Agnes Moorehead and Orson Welles impersonating everyone from Winston Churchill and Franklin and Eleanor Roosevelt to Hitler, Stalin, and Mussolini. It also served as the prototype for the well-known film *March of Time*, which in 1937 would win an Academy Award for having "revolutionized" the movie newsreel.[6] The radio version left a legacy of its own. According to a chronicler of the series, "[I]t reached millions of Americans during its fourteen-year history, provided publicity for *Time*'s publications that was of incalculable value, and was consistently rated in industry-conducted polls as one of network radio's most popular dramatic programs."[7] In addition, it influenced young writers learning how to create reality-based scripts for the new medium.[8]

Another long-running series premiered on CBS in 1935. *The Cavalcade of America* dramatized notable events from the nation's past. It was sponsored by the DuPont chemical company in response to Senate hearings on war profiteering during World War I, an investigation that led to DuPont being branded a "merchant of death." As a consequence, the company prohibited *Cavalcade of America* from making any reference to war or using sound effects of gunshots or explosions. Likewise, any discussion of strikes or labor unrest was strictly forbidden. Instead, as the sometime *Cavalcade* writer Erik Barnouw recalled, the series celebrated American "explorations, inventions, and humanitarian progress," including the achievements of women (but not of African Americans, who were virtually ignored). Such progress was frequently attributed to the advancement of science, as underscored by DuPont's advertising slogan, "Better things for better living through chemistry." Ironically, the series' bowdlerization of history appealed to some pacifist radio writers, who preferred not to write about violence. Consequently, according to Barnouw, "the public relations interests of a munitions manufacturer dovetailed for the moment with the social concerns of young writers of liberal and even leftist bent."[9]

Such writers and artists were plentiful during the Depression and New Deal years. The historian William Stott has asserted that "the primary expression of thirties America" was that of documentary (broadly defined), particularly that with "an axe to grind" and aimed at "social improvement."[10] Documentary expression appeared in a host of media, including novels such as John Steinbeck's *The Grapes of Wrath;* government-sponsored art, theater, and dance, through the Works Progress Administration; the photojournalism of Margaret Bourke-White, Dorthea Lange, and Walker Evans; and the literary journalism of James Agee and Walker Evans's *Let Us Now Praise Famous Men*. Documentary film also came to the forefront, following the example of the British filmmaker John Grierson, who declared that he viewed "cinema as a pulpit" devoted to promoting "organized and harmonious living."[11] In America, according to one historian, early documentary film "was sharply political in origin" and frequently championed organized labor. Then came "a more moderate reflection of the liberalizing change sweeping in the nation," with government-commissioned films such as Pare Lorentz's *The Plow That Broke the Plains* (1936) and *The Flood* (1937), promoting New Deal agriculture and conservation programs.[12]

Radio soon proved to be a uniquely powerful documentary vehicle. In Stott's words, it was "the ideal medium for putting the audience in another man's shoes [through] the immediacy of the human voice."[13] Similarly, the radio historian Bruce Lenthall argues that listeners "formed imagined but meaningful relationships with radio voices" in the 1930s, raising hopes that the medium "had the capacity to inform listeners and involve them directly in national civic affairs" to the extent that "radio might reinvent democracy."[14] Radio attracted writers such as Arch Oboler, William Robson, Arthur Laurents, Arthur Miller, Langston Hughes, Orson Welles, and Norman Rosten. "'The writer of the 1930s and 1940s belonged to a generation in rebellion,'" Rosten later said. "'He was an accuser [whose] enemy was poverty, war, fascism, corruption in high places or low.'"[15]

Such rebellious passions, although by no means the norm on network radio, did occasionally find a place to be heard. The failed radio-reform movement of the early 1930s at least encouraged the networks to demonstrate a commitment to public service, if for no other reason than to try to discourage future attempts at more stringent regulation. In the case of CBS, they also presented an opportunity for the network to distinguish itself from its more powerful rival, NBC. Under William Paley, CBS became an innovator in its "sustaining" or noncommercially sponsored programming, which as of the 1930s constituted two-thirds of its schedule. The network attracted new affiliates by offering such shows for free.[16] Sustaining programming was ideal for documentaries and other programs that were not obligated to please sponsors by attracting large audiences, which offered young writers the advantage of working without the restrictions imposed by a *Time* or DuPont. So it was that in 1937, CBS's sustaining *Columbia Workshop* presented Archibald MacLeish's verse play *The Fall of the City*, a thinly veiled antifascist allegory that "seemed to go to the heart of the terror of its time," as a historian later put it—namely, the terror associated with the Spanish Civil War and the rise of Hitler and Mussolini.[17] MacLeish enthused about radio's potential of reaching "an infinitely greater number of people" than live theater could, and the success of *The Fall of the City* encouraged the production of similar works.[18]

At the same time that radio docudrama was beginning to flower, radio journalism also prospered, having overcome stiff resistance from newspapers and wire services that had unsuccessfully fought to limit the kind and amount of news that radio could broadcast.[19] Paul White and Ed Klauber at CBS and Abe Schechter at NBC had begun news operations at their respective networks by the early 1930s, followed by G. W. "Johnny" Johnstone doing the same at the Mutual radio network.[20] In contrast to the accusatory advocacy of the docudramatists, the network news organizations aspired toward balance and objectivity. Klauber mandated that those "presenting or analyzing the news must not express their own feelings" and that an "unexcited demeanor at the microphone must be maintained at all times." Once Edward R. Murrow assumed oversight over CBS's European coverage in 1937 and started assembling the group of correspondents who would become known as the Murrow Boys, he sought to recruit those "'who would be steady, reliable, and restrained,'" as he later put it.[21]

After the London Blitz erupted in 1940, the line between objectivity and advocacy became increasingly blurred for Murrow. Radio had extensively covered the buildup to the war, with CBS producing its first European news roundup in conjunction with the Nazi occupation of Austria in March 1938 and all the networks devoting saturation coverage later that year to the Munich crisis and subsequent German takeover of Czechoslovakia. Within a year after Hitler's invasion of Poland in September 1939—and again as reported by radio—the Nazis controlled the majority of Europe, with Britain standing virtually alone against them. Murrow's subsequent broadcasts from London became legendary. Although nominally remaining "steady, reliable, and restrained" and never openly urging American intervention on behalf of the British, Murrow acknowledged to his wife that he wanted to shake up listeners and "let them have it": "Now I think is the time. A thousand years of history and civilization are being smashed." In a letter to his parents, he declared that he was "trying to talk as I would have talked were I a preacher" and that radio provided "a powerful pulpit," much as John Grierson would say of film.[22]

Those broadcasts prompted Archibald MacLeish to lavish praise upon Murrow at a testimonial dinner back in New York in 1941: "You burned the city of London in our homes, and we felt the flames that burned it. You laid the dead of London at our doors, and we knew the dead were our dead." For his part, Murrow told the assemblage that although it was "no part of a reporter's function to advocate policy," the English believed "that unless the United States enters this war, Britain may perish." He added that it could no longer be doubted that "the important decision, perhaps the final decision, that will determine the course of human affairs will be made not in front of Moscow, not on the sands of Libya, but along the banks of the Potomac."[23] Five days later, the Japanese attacked Pearl Harbor.

Wartime Documentary and News

With America now in the war, radio journalists were less apt to maintain an objective façade. "'This is a war for the preservation of democracy,'" Paul White told CBS News personnel in a memorandum. "'The American people must not only always be kept vividly aware of this objective, but of the value of every man, woman, and child in the nation of preserving democracy.'"[24] The journalists sent back appropriately vivid accounts of the war's progress in Europe and the Pacific. Murrow was far from the only such journalist and perhaps received too much attention in comparison with some of his peers at CBS and the other networks.[25] Still, he would provide some of the most celebrated examples of radio war reportage, including a 1943 nighttime airstrike against Berlin ("the small incendiaries were going down like a fistful of white rice thrown on a piece of black velvet"), a paratroop drop over Holland the following year ("they seem to be completely relaxed, like nothing so much as khaki dolls hanging beneath a green lampshade"), and the

1945 liberation of Buchenwald ("It appeared that most of the men and boys had died of starvation; they had not been executed. But the manner of death seemed unimportant. *Murder* had been done at Buchenwald").[26]

Wartime documentarians embraced even more fervently what one historian has called the role of "bugler" in producing work that would serve as an "adjunct to military action, [a] weapon of war."[27] Hollywood's leading directors were enlisted in that mission. Some produced films about courage and heroism on the front lines, as with William Wyler's *Memphis Belle*, John Ford's *The Battle of Midway*, and John Huston's comparatively bleak *The Battle of San Pietro*. The *Why We Fight* film series overseen by Frank Capra aimed more at explaining the rationale behind the war, as the title implied. The films in the series avoided dwelling on the more unpleasant realities of American life, with *The Negro Soldier* failing to mention segregation. At the same time, they boosted some causes dear to those on the political left, with *Prelude to War* taking an anti-Franco tack in its recounting of the Spanish Civil War.[28]

Idealistic radio dramatists threw themselves behind the cause, escaping what Erik Barnouw recalled as a "repressed and frustrated period" before America's entry into the war, during which "radio became increasingly nervous about unneutral drama, fearing the charge of warmongering."[29] With such concerns now past, Norman Rosten wrote a paean to the Soviet army, which was fighting on the Allied side at the time. Morton Wishengrad's *The Battle of the Warsaw Ghetto* commemorated the Polish Jews who had died resisting the Nazis. Arthur Laurents's *The Last Day of the War* looked at the readjustment of disabled veterans returning home.[30] Domestic tensions received more attention than that provided by the likes of the *Why We Fight* films, although the emphasis was still on promoting a unified home front. Arch Oboler wrote a radio play around the song "The House I Live In" that celebrated the country's religious and ethnic diversity.[31] Station WMCA in New York presented "The Negro Domestic" as part of its *New World A-Coming* series. "If there is one thing that irritates Negroes today—it is the 'mammy' legend often romanticized in song and story," the program began. "Mammy—the epitome of the patient, contented slave—doesn't live here anymore!"[32] "Japanese Americans" commissioned by the Armed Forces Radio Service for the series *They Call Me Joe* stressed that Japanese American soldiers "fight in the same cause as all Americans."[33]

Especially notable was William Robson's *Open Letter on Race Hatred*, which aired on CBS in 1943 a few weeks after wartime riots in Detroit had killed dozens, the majority of them African American. Robson later wrote that the scheduled broadcast was postponed twice until it was felt that the program would successfully "throw the light of truth on the Detroit incident, without inciting either whites or Negroes to riot elsewhere."[34] CBS transmitted a recording of the program to its affiliates over closed circuit prior to the national broadcast so that the stations could decide for themselves whether to air it; ninety-six ended up doing so, including some in the South. In the program's climax, a narrator expressed the hope that the show had impressed upon listeners "the irreparable damage race hatred has already done to our prestige, our war effort, and our self-respect," as well as the determination never to "allow intolerance or prejudice of any kind to make you forget that you are first of all an American with sacred obligations to every one of your fellow citizens."[35]

Programs such as *Open Letter on Race Hatred* have since been criticized for "limiting the ability of blacks to speak for themselves."[36] (Indeed, Langston Hughes declared in the 1940s that American radio was "almost as far from being a free medium of expression for Negro writers as Hitler's air[waves] are for the Jews."[37]) At the time, though, such programs were commended for representing radio at its best. *Time* pronounced *Open Letter on Race Hatred* "one of the most eloquent and outspoken programs in radio history," whereas the *New York Times* praised CBS for having had the courage to air it, especially considering the networks' usual timidity regarding controversial issues: "This was radio in the public interest. It does not happen every day."[38]

Radio at the End of the War

At the end of World War II, radio was at the height of its power. An observer at the time called the war "radio's golden period," literally so in terms of the wealth broadcasters amassed.[39] Due to tax laws discouraging war profiteering, companies found it prudent to spend their profits on advertising, and due to a wartime shortage of newsprint, advertisers turned to radio instead. The result was skyrocketing income for broadcasters. Beyond money considerations, "World War II helped make radio an indispensable part of public life," as a historian has since written.[40] "Nine out of ten American families own radio sets—more than own automobiles, telephones, or bathtubs," CBS boasted to its stockholders not long after the war ended. "Today more Americans spend more time listening to radio programs than they spend doing anything else, except working and sleeping." The network also quoted a poll by the National Opinion Research Center that said 67 percent of those surveyed thought radio had been the medium that had best served the public during the war, compared with 17 percent who felt that newspapers had done the best job.[41]

As for radio news, the time immediately after the war would represent its own "Golden Age," as recalled by the journalist Edward Bliss: "For correspondents who worked in places like London, Paris, and Rome, these were the last halcyon years.... They lived on expense accounts in luxury apartments to which they invited leaders of government for dinner."[42] CBS reported that it had devoted more than 7,500 different broadcasts to news or sports programming during 1945 alone, including those "from twenty-nine places in Europe, from twenty-three in the Pacific, and from sixteen other points outside the United States."[43] NBC similarly bragged that its Radio City headquarters "kept in constant touch with sixty-five NBC reporters and commentators at strategic posts in every quarter of the globe" who provided at "frequent intervals each day up-to-the-minute news reports and analyses from the places where history was being made."[44]

The networks vowed to build upon what they said they had already accomplished. "The NBC staff, both at home and abroad, is just coming into a period when experience, balance, and news sense will enable it to achieve heights never permitted during the war," said the NBC news director William Brooks, while also reclaiming the mantle of objectivity: "We must not attempt to preach or crusade for causes. Ours is the job to report, and to report accurately and impartially."[45] NBC placed a new emphasis on domestic news in its program *News of the World*, which aired each weeknight. Similarly, CBS in 1946 launched a daily program called *News of America* that paralleled the network's *World News Roundup*; the new program featured reports from CBS affiliates across the country. Local stations began investing in their news operations, with more than three-quarters of them airing more local news by 1946 than they ever had previously, according to one survey.[46]

Networks and stations also aired public-affairs series in addition to their daily newscasts. Examples included CBS's *Assignment Home,* which, according to the network, "probed the many personal, social, and economic problems encountered by the millions of newly returned Americans in their efforts at readjustment"; *Cross Section—USA,* which enlisted the cooperation of nine labor, business, and agricultural organizations in discussing economic issues; and *Frontiers of Science,* which explored "the lives and achievements of research scientists as contributors to modern culture."[47] NBC featured series such as *Doctors at Home* (discussing "questions of hygiene and medicine in the postwar community"), *They Came This Way* (relating "the fight for freedom throughout human history"), and *Our Foreign Policy* (providing "an informative weekly discussion of the position and purposes of the United States in the world family of nations").[48] Local efforts ranged from *New World A-Coming* on WMCA in New York (which had aired "The Negro Domestic") to the Boise station KIDO's *Operation Idaho* on the state's business and tourism climate.[49]

Radio commentators enjoyed national attention—if not notoriety—while presenting a comparatively diverse range of opinion. In what it dubbed a "Capsule Appraisal of Radio's Know-It-Alls," *Variety* in 1945 listed thirty regular commentators on domestic and global affairs. They included CBS's Murrow and his "boys" Charles Collingwood and William Shirer. *Variety* identified them as either "middle of the road liberal" or (in Collingwood's case) just "middle of the roader." Other network analysts said to lean toward a liberal viewpoint included Cecil Brown, who had moved to Mutual after quitting CBS in a wartime dispute over censorship; Mutual's Raymond Swing; NBC's Robert St. John and John Vandercook; and Walter Winchell, who had been a staunch supporter of Franklin Roosevelt's New Deal. Among the prominent names whom *Variety* listed as being on the opposite side were NBC's H. V. Kaltenborn and Lowell Thomas (each labeled as "conservative"), plus Mutual's Upton Close and Fulton Lewis Jr. (each labeled as "extreme reactionary"). NBC's Drew Pearson was identified as being "changeable" in his views, whereas Mutual's Gabriel Heatter was described as being merely "confused."[50]

With all the attention devoted toward radio and its programs and personalities, the fledgling medium of television paled in comparison. CBS had had a television news department since 1941, but it was "ignored, dismissed, or ridiculed" by radio veterans such as Murrow and his cohorts.[51] "As a broadcaster and not a manufacturer of receiving sets, CBS realizes that until a full-fledged television audience is created, there can be little expectation of the income necessary to put television on a self-supporting basis," CBS told its stockholders in early 1946. Nonetheless, the network asserted that its proposed new color-TV system, broadcasting in ultra-high frequencies, would create such a robust audience.[52] Its rival NBC was in fact "a manufacturer of receiving sets" via its parent corporation RCA, and it declared that it was rapidly developing regional TV networks on the East Coast with an eye toward eventually connecting the entire nation via coaxial cable. "We no longer are required to predicate plans for television on the winning of the war," NBC President Niles Trammell told the FCC in October 1945. "Victory has been won. Peace is here. Television is ready to go."[53]

Thus there already were signs that radio's days as America's preeminent broadcast medium might be numbered. There also were signs that a backlash might be brewing against perceived liberalism on radio. As early as 1943, the commentator Quincy Howe had written of a movement against left-leaning commentators, and that movement would rapidly accelerate after the war with at least one commentator heretofore seen as liberal—Walter Winchell—growing more reactionary.[54] In the fall of 1945, the House Committee on Un-American Activities (HUAC) began investigating alleged subversion among radio commentators. Liberals were not easily intimidated, however. *Variety* approvingly reported that "showpeople in and out of radio organized a campaign to deflate the committee's smearing tactics."[55] The campaign was called the "Mobilization against Thought Police in the USA" and was backed by members of the AFL and CIO labor organizations. For a time, such efforts were seen as having the potential to dismantle HUAC altogether.[56]

In fact, despite the horrors of the war just ended and the challenges and uncertainties that lay ahead—with much of the world in ruins, relations with Russia growing ever more strained, and terrifying new weapons of mass destruction at the ready—many Americans, liberal or otherwise, were filled with an almost giddy optimism. Harold Russell had lost his hands during the war and would win an Academy Award for his role as a battle-scarred veteran in the 1946 film *The Best Years of Our Lives*. "The guys who came out of World War II were idealistic," he said years later. "They sincerely believed that this time they were coming home to build a new world."[57] Radio was seen as a potential means toward that end. "Radio must realize that the future of our democracy—the only form of government in which free radio can survive—depends on an educated and well-informed America," wrote two commentators in 1945. "Radio must realize that it can be the instrument for accomplishing this. It has the power; it has the audience. It remains to be seen if it has the will."[58]

Others expressed skepticism that radio, with its "lowest-common-denominator philosophy of programming," would ever exert that will without considerable prodding.[59] For all the ballyhooed news and public-service programming that had been broadcast during the war, radio remained foremost a vehicle for commercial gain. "American radio uses [its] power mainly for merchandising," Erik Barnouw wrote in 1945. "Drugs, foods, and tobaccos, chief financiers of the medium, fill many of the choicest listening periods with gag comedy and escape drama."[60] One critic asserted that it would depend on public "interest, indignation, and participation" to pressure the FCC and Congress into forcing radio to rise above "the kindergarten class of most movies and comic publications which characterize the American scene today."[61]

For his part, the FCC Chairman Paul Porter declared that "radio must grow up" and urged listeners to demand better things from broadcasters: "The air waves do not belong to the Government, or to the FCC, or to broadcasting stations. They belong, by law, to you—the public."[62] The FCC by 1945 had begun preparing its Blue Book report aimed at raising radio's standards, in many ways paralleling the contemporaneous work of the Commission on Freedom of the Press, which would issue recommendations for more socially responsible journalism.[63] It was a heady time for those who dreamed of a new and improved postwar society. One historian writes that "the dawn of the Atomic Age marked a watershed moment when major institutions were open to reformist impulses, and power relations, both national and geopolitical, were in flux."[64]

* * *

This, then, was the backdrop against which Edward R. Murrow was said to have enthusiastically urged his colleagues to show Americans what radio could accomplish in peacetime. Murrow was returning home to assume command of CBS's news and information programming, and for a time journalists with their stated ethic of objectivity and dramatists with their commitment to social change would find common cause before their paths diverged. "Reformist impulses" acting within and upon radio would compel it to produce more adventuresome programming, even as a rightward shift in the country along with broadcasting's ingrained tendency toward amassing profit (including through the potential gold mine of television) would start undercutting those impulses.

At that portentous moment, the man who arguably was America's preeminent radio dramatist was spelling out his own credo for broadcasting. "I believe radio has a higher destiny than merely to sell soup and soap," he wrote in 1945. "I think it's an art apart; a social force which can figure vitally in the keeping of the peace and the making of a clean and orderly world."[65] His name was Norman Corwin, and he was about to undertake one of the most ambitious radio documentary projects ever attempted.

2

One World

Norman Corwin was born in 1910 in Boston.[1] As a young man, he worked as a newspaper journalist and movie-studio publicist, but he always viewed radio with wonder—it was "'a theater without walls whose roof was the sky itself, a theater not for the optical but the mind's eye, a locus in which thought, language, imagery, and metaphor enjoy an authority seldom honored in stage or film productions.'"[2] Corwin began writing for radio and joined CBS in 1938. His timing was propitious. "Because CBS was a young network at that time, and it was pitted against a network whose pyramidal base was RCA and [that] had all the money in the world," he recalled, "they gave me great freedom, and I must say I prospered from that."[3]

CBS's reputation for encouraging experimentation attracted others who, like Corwin, were progressively minded. Working at the network was compared to being "at the top of a liberal mountain."[4] Corwin wrote all manner of shows, including one a week for the 1941 series *Twenty-six by Corwin* that included comedy, fantasy, and melodrama. However, he became best known for what he called "murals and polemics"[5] that expansively confronted the great issues of history and the present moment. Inevitably, those included the rise of fascism. One of his earliest CBS efforts, 1939's *They Fly through the Air with the Greatest of Ease,* was a far cry from the idealized niceties of programs such as *Cavalcade of America.* It told of a bomber crew of unnamed nationality that nonchalantly killed civilians Guernica-style before meeting a bloody end themselves. Corwin said that he wrote it in anger at Spain's "deliverance into the hands of the loathsome Franco through the active assistance of Hitler and Mussolini and the passive assistance of practically everybody else." Despite the controversial subject matter, CBS made no effort to censor the program and even repeated it twice on the air.[6]

Soon enough, the United States itself was at war. The government asked Corwin to write a program commemorating the sesquicentennial of the Bill of Rights, which coincidentally fell on December 15, 1941, the week after Pearl Harbor. Airing on every network and ending with a live address by President Roosevelt, the show celebrated American freedoms and boosted the nascent war effort. Corwin continued in that vein with programs such as *This Is War!*, which excoriated the Axis powers, and *An American in England*, which sent Corwin to London to work with Edward R. Murrow as producer.[7]

Wartime radio programs have been criticized as blatant propaganda not only for the U.S. military effort but also for big business.[8] At the time, though, Corwin had no qualms about engaging in what he called the "indoctrination" of listeners—too much was at stake. "Americans want to know what kind of world there will be at the end of this war," he said in 1942, adding that "the peace will take care of itself" if the people were "taught the nature of the democracy they live in and the nature of the Fascism against which they are in mortal combat."[9] The culmination of Corwin's efforts toward that end aired on the night of the Nazi surrender in May 1945. *On a Note of Triumph* opened with a fanfare composed by Bernard Hermann, followed by some of the most famous lines in radio history: "So they've given up. . . . Take a bow, G.I.! Take a bow, little guy! The superman of tomorrow lies at the feet of you common men of this afternoon."[10]

The program cemented Corwin's reputation as being what *Time* called CBS's "boy wonder" and master of "his favorite stock in trade: the supremacy of the common man."[11] As his fame grew, he attracted criticism. A lengthy *New Yorker* profile needled him for his bachelor habits and his perceived hypochondria and vanity ("Norman enthralls [his friends] with ear-shattering recordings of his playlets; when one is finished, another is placed upon the turntable"). The profile was in the portentously mock form of just such a Corwin radio play, complete with interjections from a "Chorus of One Hundred Little Guys from Everywhere in the World."[12] Even otherwise flattering stories about Corwin opined that he could be "mawkishly patronizing about the little people" who were his subjects.[13]

In response, Corwin said that the "little guy" was not a "milquetoast figure." Instead, he was "literally little in income, property, investments, social standing, influence, and renown," as well as fundamentally "peaceful, uncomplaining, good-hearted. . . . That's what makes him little, and that's why I'm for him."[14] Corwin's sentiments were thus very much of a piece with the socially minded fiction and journalism of the Depression years, Ernie Pyle's wartime dispatches about the selflessly brave infantryman, and the composer Aaron Copland's *Fanfare for the Common Man*, which had its debut performance during the war.[15] Corwin later expressed his debt to progressively minded poets such as Walt Whitman and Carl Sandburg (who would praise *On a Note of Triumph* as "one of the all-time great American poems"), as well as his admiration for Archibald MacLeish and Stephen Vincent Benét (both of whom had written antifascist plays for radio).[16]

Reviewing the Bill of Rights program in 1941, *Variety* also detected an affinity between Corwin's work and that of the film director Frank Capra.[17] However, if Capra's pronouncements could sound simpleminded ("people's instincts are good, never bad"), Corwin was what the critic Gilbert Seldes called "a mature anti-fascist" aligned with the New Deal and the wartime liberalism of Henry Wallace.[18] "Some have spoken of the 'American Century,'" said Vice President Wallace in 1942, countering the *Time* publisher Henry Luce's vision of American global preeminence. "I say that the century on which we are entering—the century which will come out of this war—can and must be the century of the common man."[19] According to one of his biographers, Wallace saw the war "as a struggle between human reason and intolerable evil" represented by fascism.[20]

Following the death of Roosevelt, Corwin saw Wallace as "'the last and best bulwark against fascism in [America and] the best guarantee of survival of civil liberties and true democratic freedom.'"[21] That political sensibility shaped Corwin's work. In *On a Note of Triumph*, he scorned "the Century of the Uncommon Aryan" to which the Nazis had aspired as well as the American isolationist press that could "lie with a straight face seven days a week and be as filthy and Fascist as a handout in Berlin." He added, "We've learned that those most concerned with saving the world from Communism usually turn up making it safe for Fascism." In contrast to those who saw such writing as pure propaganda, Corwin saw it as a blend of art and conscience, which to him went together "as readily as gin and tonic."[22]

Conscience also compelled him toward another tenet of Wallace-style liberalism: that the war was "a cauldron out of which the greater good of social reconstruction must come," leading to a united global society not dominated by any one power.[23] Corwin was an optimist who declared that there was "nothing to be said for cynicism and despair, and everything to be said for getting out and working toward a better world." Achieving that world, though, would require concentrated labor to overcome the inevitable obstacles. "You can win a war today and lose a peace tomorrow," warned the narrator in *On a Note of Triumph*, before closing the program with a solemn petition:

> Lord God of test tube and blueprint . . .
> Post proofs that brotherhood is not so wild a dream as those who profit by postponing it pretend:
> Sit at the treaty table and convoy the hopes of little peoples through expected straits,
> And press into the final seal a sign that peace will come for longer than posterities can see ahead,
> That man unto his fellow man shall be a friend forever.[24]

The following year, Corwin would be able to assess the prospects for global friendship firsthand. He did so in the name of another liberal politician, albeit one who had once campaigned against both Henry Wallace and the New Deal.

Willkie's "One World"

Wendell Willkie was the dark-horse Republican nominee for president in 1940, losing to the Roosevelt-Wallace ticket.[25] Despite his defeat, Willkie strongly supported controversial Roosevelt measures such as the Lend-Lease program to aid Britain prior to America's entry into the war. As a result (and possibly as a way of ensuring that Willkie would not be around to campaign for Republican congressional candidates), Roosevelt dispatched him on a fact-finding world tour at the height of the war in 1942. Willkie met with Joseph Stalin in Russia and with Chiang Kai-shek and Chou En-lai in China. He also visited Africa and the Middle East, talking to the British General Bernard Montgomery and the French leader Charles de Gaulle as well as to U.S. military leaders. Willkie's book about his travels, *One World*, quickly became a bestseller.

Willkie alienated many fellow Republicans with his newfound ties to Roosevelt as well as his diehard opposition to isolationism. He argued in *One World* that instead of resorting to "narrow nationalism" or "international imperialism," America after the war had to cooperate in forging "a world in which there shall be an equality of opportunity for every race and every nation." Regarding racial equality at home, Willkie was arguably more progressive than Roosevelt: "When we talk of freedom and opportunity for all nations, the mocking paradoxes in our own society become so clear they can no longer be ignored. . . . [W]e must mean freedom for everyone inside our frontiers as well as outside."[26] One scholar has argued that Willkie's philosophy of "wise internationalism abroad" combined with "a domestic racial politics of constructive difference" could be compared to that of multiculturalism.[27]

Willkie unsuccessfully campaigned for the 1944 Republican presidential nomination before succumbing to heart failure that October. To honor his memory, the Common Council for American Unity and Freedom House (an organization Willkie had helped found) established the One World Award to subsidize a trip similar to Willkie's. Norman Corwin was named the 1946 winner for his "contributions to the concept of One World in the field of mass communication." Corwin's acceptance was contingent upon CBS's approval for him to take the trip and make it the subject of a potential radio series. William Paley readily agreed, and CBS paid for an engineer and equipment to accompany Corwin on his journey.[28]

The award was presented in New York City in February 1946. The following month in Beverly Hills, Corwin was honored with readings of his work by Paul Robeson, among others. Already the prospects for peace looked bleak. The public had not known of the atomic bomb when *On a Note of Triumph* aired the previous spring. In addition, Winston Churchill was now calling for a U.S.–British alliance against the Soviet "Iron Curtain" in Europe. Many condemned Churchill's broadside against America's erstwhile ally, but it foreshadowed a shift in press and public opinion that would help fuel the cold war.[29]

Corwin addressed those subjects in speeches in New York and Beverly Hills. "If oneness of the world is a dream, then we are proud to call ourselves dreamers," he said, adding that when the world absorbed "the logic and truth" of Willkie's philosophy, "we shall be able to dismiss the fear of being vaporized suddenly and without warning." As for Winston Churchill (who had been voted out of office the previous year), Corwin declared, "Willkie passed from defeat into greatness. Churchill passed from greatness into defeat." The key issue was "whether now that we've reached the Atomic Age, we shall have One World or Two."[30]

Corwin was not alone in such views. Critical media treatments concerning the effects of atomic weapons and the need for global amity were common in 1946, the most famous being John Hersey's *Hiroshima*.[31] ABC presented a condensed four-part reading of Hersey's story immediately after the *New Yorker* published it. "This chronicle of suffering and destruction is not presented in defense of an enemy," an announcer said before the first segment. "It is broadcast as a warning that what happened to the people of Hiroshima a year ago could next happen anywhere."[32] In a similar vein, CBS aired *Operation Crossroads* in anticipation of that summer's Bikini atomic-bomb tests. It concluded with a call for international control of the bomb at a time when the United States had sole possession of the weapon.[33]

Perhaps most memorable was *Unhappy Birthday*, which marked the first anniversary of the Hiroshima attack. ABC presented it in cooperation with the group Americans United for World Government. The program began by dramatizing the presumed effects of an atomic attack on a young boy. (Whimpering child: "Mommy! Daddy! My arm!" Mother: "His arm! His left arm is gone! And his right arm is slowly—*aaaaaagh!*") There followed a cautionary tale about "Joe Mason, electrician and veteran," stubbornly ignoring warnings about the atomic threat from a real-life cast including the writers Fannie Hurst and Merle Miller and U.S. Senator Charles Tobey. "Joe" dismisses them all as alarmists: "Atomic war? World government? Sounds like a lot of propaganda. Yeah, that's all it is. Just a lot of propaganda!" Then he experiences an atomic-fueled nightmare, the same one that had been heard at the start: his son loses his arms and his family is marched off to a slave camp. An announcer at the conclusion told the audience that it was not too late to realize that "other nations will soon have bombs of their own" and that "therefore, we must have world law to guarantee world security. Those are the facts. Learn them. Remember them. Tell them to others. Keep alert. Keep thinking!"[34]

Against that apprehensive backdrop, Corwin prepared to depart on his global travels. Not only were there growing tensions with the Soviet Union amid the specter of atomic annihilation, but there also was ongoing civil warfare in some of the countries he was about to visit amid the vast destruction wreaked by the war just ended. Corwin told his Beverly Hills audience that he knew that he would find "'turmoil and trouble, hunger, poverty, and restiveness.'" Yet he also expected to find "'hope intermingled everywhere'" and "'people who are eager to join with you in the making of a better day.'"[35]

Using Recordings

The hope for a better day was at the heart of Corwin's mission overseas. He later said that he felt that journalists had done a fine job of reporting the bloodshed and turmoil of the war. In contrast, he believed "that I could honor the concept of One World if I concentrated on those areas of concord, of healing. And in the interviews I had with people in high and low stations, that's what I emphasized."[36] It was important to Corwin to talk not only with government officials and others in high places but also with common people on the street.

Finally, he wanted to record the interviews and so-called natural sound from the various locales to use on the air rather than follow the established course of re-creating everything in the studio with actors and sound effects, as, for example, *The March of Time* had done. During the war, Corwin had written much of a government-sponsored series for CBS called *Passport for Adams* that starred Robert Young as a journalist sent on a worldwide trip to visit and report on America's allies, including Russia. The series had dramatized actual incidents that various war correspondents had experienced. For *One World Flight*, Corwin set out to "have people in their own countries and their own accents present their own views," feeling that "dramatizations would not be true to the essence" of the trip.[37]

The two most powerful networks, CBS and NBC, had long banned the broadcast of recordings to protect the networks' unique status as nationwide purveyors of live news and entertainment. As one observer said at the time, they viewed recordings "as a threat to their very existence. If programs are mailed out to stations in the form of records, they ask, what is the value of a network?"[38] There also were concerns that recordings would give advertisers an excuse to buy time on only selected network affiliates while bypassing others entirely. Occasional exceptions were made to the recording ban, as with Herbert Morrison's famous report on the 1937 Hindenburg disaster and George Hicks's on-scene account of the Normandy landings on D-Day in 1944, but the CBS and NBC ban still largely held firm as of the end of the war.[39]

There was another obstacle: the technology for making field recordings outside the studio was crude and bulky. John and Alan Lomax had taken a sound-recording truck on the road and produced audio documentaries for the Library of Congress, including 1941's *Mister Ledford and the TVA*, which featured the actual voice of a Georgia farmer, Paul Ledford, as he pondered having to leave his home to make way for a new dam. It aired on approximately sixty radio stations and was especially well received in the South.[40] Such programs followed the lead of the BBC overseas in using recordings of real-life people. The sound trucks, though, were large, slow, and not easily maneuverable. One used by the BBC weighed seven tons and was twenty-seven feet long, making it impossible to navigate country roads.[41] Hand-carried machines that recorded on shellac or glass discs were little more practical; a Library of Congress expedition to South America used equipment that required hiring "six natives" just to haul it. There also were constant problems with finding a reliable power supply and with the fragile discs breaking.[42]

However, by 1946 Bing Crosby was recording his weekly radio show in the studio, allowing him to perfect the broadcasts and avoid having to perform the same show live twice in a row (first to the eastern half of the country and then to the western half). When NBC balked at using the recordings, Crosby moved to ABC, which, like Mutual, had no recording ban. Other radio stars, such as Al Jolson, were intrigued by Crosby's idea, and some followed Crosby to ABC, which also soon started adding disc jockeys to its schedule. Such developments increased pressure on CBS and NBC to relax its ban. Thus CBS head William Paley allowed Corwin to make recordings overseas and put them on the air.[43]

Unfortunately for Corwin, plastic audiotape and the magnetic tape recorder would not come into widespread use until a couple of years after his trip. Instead, he and his production engineer, Lee Bland, would go abroad with a General Electric wire recorder—a machine that recorded onto long spools of steel wire not much thicker than a human hair. The military had used wire recorders during the war, and Fred Friendly in his role as an information officer had captured the sounds of bombing runs, prompting him to say that such machines "'would transform the way we do journalism.'"[44] One had been used to record Edward R. Murrow's famous broadcast of the Allied paratroop drop over Holland. The wire recorder was far inferior to audiotape in terms of sound fidelity and ease of editing, but in Corwin's hands as of 1946, it represented the most ambitious use that radio had ever made of field-recording technology.[45]

The Flight

Four months of preparation were required for a trip that would cover four months and thirty-seven thousand miles. Corwin and Bland required sixty visas between them. In response to Swedish inquiries concerning his race and religion, Corwin puckishly wrote "Human" and "Monotheism." As for a Soviet question regarding his political affiliation, he replied "Republican and Democrat" after resisting an initial urge to respond with "none of your business." "I could not afford to risk losing the visa by offending them on a point of their peculiar xenophobia," Corwin later recalled, "because the whole *raison d'être* of a One World Flight was to visit *both* existing worlds."[46]

Freedom House wrote Dwight Eisenhower requesting War Department accreditation for Corwin, which was necessary for him to visit Tokyo, then occupied by the American military. Although the State Department at first said that Japan should be removed from the itinerary while also raising concerns about the Soviet Union, Corwin was granted access to both places. Foreign consulates were contacted for information and suggestions regarding currency, immunizations, weather conditions, and what and whom to record. CBS also enlisted the support of its overseas correspondents, including Howard K. Smith in Britain, Bill Costello in China, and George Polk in Egypt. Wire spools for the recorder had to be shipped in advance to each country due to airplane weight and space restrictions. Even so, Corwin and Bland would lug almost four hundred pounds of luggage around the world, with the recorder and its accessories taking up more than half that weight. The final itinerary, to be reached via a motley collection of commercial and military aircraft, included London, Paris, Copenhagen, Oslo, Stockholm, Warsaw, Moscow, Prague, Rome, Cairo, Delhi, Calcutta, Chungking, Shanghai, Tokyo, Manila, Sydney, and Auckland.[47]

Corwin and Bland departed from New York for London on June 15, 1946. President Truman sent a congratulatory letter that said that "anything which can promote better relations in a world too long divided by suspicion and ill will is of value to our time."[48] The CBS program chief Davidson Taylor wished them farewell, announcing that the trip would result in the first of a series of programs in which "dramatist-correspondents go to various parts of the world and come back and tell the people of the United States via radio programs what they have seen of our friends and the people we would like to be our friends."[49] Taylor and Edward R. Murrow had recently been made network vice presidents, and *Variety* reported that the executives intended to eliminate "the universal stigma that applies to sustaining programs" by raising them to "the same high level of commercial programs." That was seen as part of a broader CBS strategy to build up its programming as a competitive response to NBC, which still dominated the ratings.[50]

Bland had been specially trained to operate the wire recorder, and a CBS engineer had said that "it is not expected that he will have much, if any, trouble" with it.[51] In fact, it began periodically malfunctioning soon after he and Corwin arrived in London, and several interviews failed to record. The same was true in France, prompting Bland to take the machine to U.S. military engineers in Germany in a failed attempt to fix it. Corwin would remember the recorder as a "monstrous device" and a "horror box" that "balked with an almost intelligent perverseness." It was heavy; the recording quality was poor, with a periodic "obliterating hum"; the wire snagged easily; and breaks and splices had to be knotted and then fused with a lit cigarette: "What could be more primitive than that in sound technology, except beating on a drum in [the] forest?"[52] The problems finally ceased after radio officials in Stockholm granted Corwin the use of their own wire recorder for the rest of the trip in exchange for the obstreperous device he and Bland had been forced to carry to that point.

In the meantime, he was doing his best to accomplish what he had set out to do. In Britain, he interviewed the writer J. B. Priestley and Prime Minister Clement Attlee, the first such radio interview Attlee had granted. In France and Scandinavia, Corwin similarly recorded heads of state, party leaders, artists, scientists, and ordinary citizens regarding the outlook for One World. He found pacifist sentiments in Sweden to be especially strong: "It was almost as though the Swedes, having been without war for 140 years, were anxious to stay that way forever. They wanted no war of any description, and to them atomic war was only a refinement."[53] The grimness of the war just past and its aftermath did not fully strike home until Poland, where Corwin toured the rubble of the Warsaw ghetto after learning of a pogrom that had just occurred in the Polish town of Kielce.

Although there were also obvious privations in Russia, Corwin for the most part found the people friendly, and he was able to move freely. He had submitted a list of thirty requests to Soviet officials. All but three were granted; the others (including an interview with Stalin) were never acted upon. Corwin's experiences differed from those of the CBS Moscow correspondent Richard Hottelet, who told Corwin and Bland that the Soviets greatly restricted his work and that of his peers. After Corwin fell ill with strep throat during his Moscow stay, Bland took his place in interviewing the filmmaker Sergei Eisenstein and the composer Sergei Prokofiev. Corwin recovered in time to interview an official of Moscow radio, who said that "the greatest possible invention would be if it were possible to invent a microphone that would not allow a lie to pass. That would make ether a very good medium."[54]

The trip then continued to Prague, which turned out to be one of the more upbeat stops. Corwin felt that Czechoslovakia was achieving a fruitful rapprochement between East and West. However, the outlook was gloomier in Italy,[55] where food and work were scarce, and gloomier still on the long journey across the Middle East and Asia. A planned visit to Jerusalem was scuttled following a hotel bombing there. Then, after a particularly harrowing airplane landing in a storm in India, Corwin's attempts at interviews in Calcutta floundered in the face of deadly rioting between Hindus and Muslims, though he was able to interview the future prime minister Pandit Nehru in New Delhi. China was also embroiled in civil war, and Corwin was unable to visit communist-controlled sections of the country, although he did conduct interviews at the headquarters of a U.S.-led effort to mediate between the two warring factions. There, despite protests from the U.S. representative present, the communist representative told Corwin that America had to cease its one-sided support of Chiang Kai-shek's government or else abandon its mediation efforts entirely, which in fact happened soon afterward. Corwin later described the incident as "the precise moment, and place, where the situation respecting China (hence all Asia) and the United States passed from hopefulness to hostility."[56]

Corwin had an off-the-record interview with General Douglas MacArthur in Tokyo but found himself constrained by the military occupation; as a result, he ended up not using any Japan material in the radio series.[57] In the Philippines, he again encountered widespread devastation as well as distrust of the Americans who had just relinquished control of the country. A dispute with ground personnel at the Manila airport delayed Corwin and Bland's departure for Australia.[58] Visits to Sydney and Auckland ended the trip on a happy note; Corwin's work was especially well known in Australia, and he was warmly received there.

Corwin returned to New York on October 27, 1946, with more than one hundred hours of recordings. "It will take a tall lot of digesting to synthesize and present this material properly," he had written William Paley from Egypt, adding that such digestion would have to wait until his return: "The classic definition of poetry—'emotion recalled in tranquility'—can perhaps be adjusted to hard observation. Notes and soundtrack and sights and smells and lights and color, recollected in the tranquility, if such it can be called, of the workshop at home."[59]

The Series

For the sake of timeliness, Corwin wanted *One World Flight* to begin airing as soon as possible. He had ten weeks before the premiere. With no established template for using recordings in a network radio documentary, Corwin, Lee Bland, and CBS engineers had to invent one on the fly. The network employed nearly a dozen "literary shock troops" to translate and transcribe the wire interviews from the "thirty-four languages and dialects" in which they had originally been recorded; the resulting log ran several thousand pages. The crew also transferred the interviews to magnetic paper tape (a precursor to plastic audiotape) for editing, painstakingly correcting speed problems that had occurred during the original recordings due to battery problems and variations in electrical current. It was the first time tape of any sort was used in a production of that scope. Corwin was said to be working "eighteen and twenty hours a day" on the project. While listening back to samples from the recorded interviews, he drafted the scripts on whatever was handy—ruled tablet paper, the backs of CBS office memos—and then marked corrections and revisions before handing them off to be retyped.[60]

"I never stopped to consider its relative freshness or radical character," Corwin recalled of the series years later.[61] At the time, though, he declared that it "would be one of the strangest and newest kinds of production known to radio." Corwin himself would narrate, using actors and dramatic reconstructions little if at all. His goal was to "adher[e] to reportorial objectivity" while avoiding "an austere social tract" and aiming at "the blunt and candid laying bare of inequities, shortcomings, and injustices."[62] The final series consisted of thirteen half-hour programs that aired weekly from January to April 1947.

In outlining *One World Flight,* Corwin wrote that an "imaginative composer-conductor" would help it "transcend ordinary documentary styles and enjoy a showmanship commensurate with good taste and authenticity."[63] According to *Variety,* Corwin needed a good composer for another reason: the powerful head of the American Federation of Musicians, James Petrillo, imposed a ban on the airing of music recordings from other countries. That made it difficult to air Corwin's recordings of "authentic local tunes" from his travels. Instead, a composer would "use the original material as melodic guides" in writing a new score for the series. Alexander Semmler, with whom Corwin had worked previously, composed the music and conducted it live during the broadcasts with a full orchestra in the studio. Corwin similarly read his narration live on the air, and actualities from the overseas recordings (including, as it turned out, a few snippets of music) were played back on cue from discs, which were considered more reliable than the easily breakable paper tape.[64]

The premiere on January 14 showcased the novel production techniques. "You are standing in the Maikovsky station of the Metro under the heart of Moscow," said Corwin at the start. "A waiting train signals to workmen in the subway to clear the track." There was a whistle blast, followed again by Corwin: "You are strolling along Hallam Street in London, and you are overtaken by a cockney street peddler selling cut iris, cut cauliflower, Yorkshire blue peas, and brand-new potatoes." Once again, natural sound was heard, and so it continued: snippets of Pandit Nehru in New Delhi, hammering on the streets of Manila, and women singing on a rooftop in Cairo, all intertwined with

Semmler's music and the hum of an airplane in flight. Everything, the CBS studio announcer stressed, had been recorded on scene around the world.

Meanwhile, Corwin stressed the series' theme: "I went looking for practical signs of agreement, for signs of a uniting world. I found fewer of these than we would wish for, but I also found plenty of hope." People were uncertain how to achieve One World, "but they all want it. They yearn for it. They are willing to work for it." There were exceptions: an Australian accountant who said that "the potential danger to world peace lies in the colored races"; a young woman in the Philippines who urged President Truman to "finish [off] Russia." There also was despair, as the premiere ended with the tearful voice of a woman in an Italian mountain village who had lost most of her family to the war. "This voice," said Corwin, "and the mute rubble of wasted towns and cities—these were the sounds of need: need for the hope and for the reality of a united world."[65]

In succeeding weeks, *One World Flight* retraced Corwin's journey, starting with England. He found London improved from his wartime visit, with "no signs pointing to air-raid shelters" and "no fresh bomb ruins." Still, the country was slow to recover fully: "That's the way it is with modern war—there are very few spoils left to the victor." The segment included clips from Clement Attlee: "[If] you continually think of the prevention of war, you don't get very far. You've got to think of positive peace. And that really depends on a greater understanding not just between governments, but between people." Corwin then described looking out the window while flying from England to France and thinking of those lost in the war: "There were thousands of [them] lying beyond view, with the English Channel across their chests. . . . And I wondered whether if those boys could talk, they would go for the idea of a world made one."[66]

The series next recounted Corwin's visits to France, Denmark, and Norway, all of which the Nazis had invaded and occupied. "Each was exhausted by its struggle," he reported. "Each was improving day by day. . . . In the occupied countries, among the men who had had the closest experience with the war—the Resistance men—there was never talk of a new war." Similar to what Corwin had found in Sweden, none of the other Scandinavian countries appeared interested in joining a new bloc against the Soviet Union or any other nation.[67]

However, Poland tested Corwin's optimism. From the air, Warsaw "looked like a painter's imaginary conception of catastrophe in which he had overstated his case. . . . Life was going on stoically against unbelievable hardships." A worker at a power plant told Corwin through an interpreter that "the human being" was the greatest power in the world, more so than atomic energy. Still, when Corwin visited the city's ghetto, which the Nazis had methodically blasted into rubble, he found only "dead silence—the silence of a city become a cemetery." He described the "little, human, heartbreaking detail: a shoe, part of a bureau, some woodwork, an old porcelain bathtub with shrapnel holes through it. . . . Here was the ultimate picnic ground of the professional anti-Semite." Knowing that Jews had been murdered in Kielce the day before his visit did not make him "hopeful of an early arrival at One World."[68]

For the Russia program, Corwin appeared to tread carefully. Elsewhere in his writings, he had decried the "disgusting perversions" of anti-Soviet sentiment: "No country, certainly not Germany or Japan, has been so vilified by the majority of the American press and by a large section of the church."[69] Although he did not share the views of those who saw Russia as "an unremitting enemy," neither did he agree with those who viewed it as "the only true friend of progress and a glistening new society incapable of doing any wrong." The country and its people showed "signs of hard wear in a hard war," the collective farm Corwin visited was not as "immaculate and highly scientific" as he had anticipated, and the police were suspicious to a degree "detrimental to the best relations" between nations.

Still, Corwin approvingly noted the exchanges he experienced with artists and intellectuals. A physicist said that using atomic energy only for the atom bomb was like using electricity only for the electric chair, and a newspaper editor paraphrased Gorky in criticizing those "who would start a world conflagration [just to keep themselves] warm." Sergei Eisenstein said that he did not want to visit America while it had a hostile attitude toward Russia. Corwin concluded the segment by saying that, based upon the "goodwill growing wild" that he found among the Russian people, there was "a reasonable foundation for improved relations" between the two countries, and "the global peace so dependent on this friendship might at last create the single world for which so many have perished in our days."[70]

Of Czechoslovakia, Corwin said that its citizens seemed to embrace "the middle way [that] reconciles socialism and private enterprise by running them side by side." Edvard Benes, who had been president in 1938 when his country's fate was sealed at Munich, had since returned to office. He spoke of the "great élan" of a people who were "working for a new liberty and for a new independence." Corwin's interviews with miners, though, indicated overwhelming sentiment favoring the Soviet Union and persistent worries about fascism. There were similar concerns in Italy, where Corwin found little interest in One World: "Hungry and jobless men look upon these ideas as a luxury, a dream to be afforded and indulged only on a full stomach." The Italians found themselves in the same plight they had been in when Mussolini came to power, having "completed an agonizing cycle from desperation to fascism to desperation" and wondering "what the victorious democracies are going to do about it."[71]

The leg of the trip starting with Egypt was for Corwin "the beginning of really morbid ignorance, squalor, disease." For someone who so fervently believed in the wisdom of the common man, the pervasive ignorance was particularly disheartening. Corwin was in Cairo on the day the country celebrated its nominal independence from Britain, but he still found people on the street who were unaware that Hitler was dead and that the Allies had won the war. In India, he found more of "the bleak area of humanity that we call backward—backward because of no inherent lacks, but because of economic stagnation and total absence of opportunity." Amid the poverty and the chaos and bloodshed of civil war, Corwin found a voice of calm in Pandit Nehru. "No nation, no people should be subjected to another," said Nehru. "No race should be considered an inferior race as a race." Corwin declared that such a "simple formula" would "make life more livable for three out of every four people alive on this earth today."[72]

However, his departure from India for China was a low point. He had passed corpses alongside the road to the Calcutta airport, and now he was headed toward another impoverished country at war. At that moment, Corwin said, "I felt that all of us—the whole shebang, you and me and the hundreds of millions of war-weary Europeans and Africans and Asiatics—must be closer to Mars than we are to One World." In China, he found more widespread ignorance: "The common man doesn't read because he can't read, and he can't read because there are no schools for him." The clash between communist and nationalist forces did not bolster his faith, even as the Chinese proudly told him that the One World philosophy had originated with Confucius.[73]

Corwin's next stop was the Philippines, which had just won its independence. Nevertheless, in addition to having suffered widespread destruction, "the country was unhappy. It was divided. It was restless. . . . [F]rom the standpoint of a traveler looking for signs of unity and peace, these were pits and roadblocks in the way of One World." Moreover, as a former U.S. commonwealth, the Philippines "was an entirely American problem." Corwin questioned whether "our culture, our men, and our money" could produce "unity and tranquility."[74]

Australia proved to be a tonic. Despite a few sour exceptions, such as the racist accountant who had been heard in the series premiere, Corwin found the people's outlook "unique and refreshing for its calmness, its lack of hysteria, and its general optimism." A dockhand suggested that "the great powers assure Russia that they do not want war" and that they recognize "her terrific sacrifice" in the last war. As for New Zealand, Corwin pronounced it "important out of all proportion to its size. I had not seen many places where people were free, busy, healthy, unworried, and at one with their minorities," those being the Maoris. Again, he commented favorably on a country "trying to reconcile the best features of private enterprise and socialism and to eliminate the worst of each," as exemplified by its national health system.[75]

The final program gave a summation paralleling a speech Corwin had made to the One World Award sponsors upon his return.[76] "In all too many of the countries overrun by war, people felt that America, as the only major power fortunate enough to have escaped blasted cities and severe privation, did not deeply enough understand the war," he said. In fact, while he had "expected to find and did find areas of suspicion, fear, and criticism of Russia in certain countries," he had been dismayed to find that the United States was also "suspected, disliked, and resented" by many and "known in some quarters by our worst features," such as racism and militarism. The biggest danger, said Corwin, was the "phobia created by factions who would have people everywhere believe there is no room in one world for more than one economic and social system"; such a phobia inevitably would lead to renewed war. There also were the dangers posed by widespread hunger, poverty, and disease:

> Against such troubles involving hundreds of millions of human beings, the hopeful things of the world may seem few and scattered, lonely and shivering and lost. But they're actually far from lost, far from hopeless. The majority of people we talked to felt to despair of the world today is to resign from it. To assume that human nature is committed to another war is to assume that suicide is a natural remedy for our ills. . . . The informed and articulate people of the earth . . . have a conscience and a will. The will is to be free and peaceful in a single world, to resist corruption, and to make none of the old mistakes anew.[77]

Reception and Controversy

One World Flight received a significant share of critical praise. The overseas recordings were "startlingly refreshing to a fiction-tired radio listener," wrote Jerome Lawrence in *Hollywood Quarterly*. *Billboard* called the series "a superb example of the great work radio documentaries, in the hands of an outstanding writer, can do." *Variety* praised Corwin for having "broadened the scope of documentary with actuality sequences, uniting it into a whole with brilliant, pungent commentary." The liberal newspaper *PM* suggested that the "series may contribute more to the cause of world peace and unity than anything that has yet been done—or will ever be done—in radio."[78]

Some gave the novel blend of actuality and commentary mixed reviews. Jack Gould of the *New York Times* wrote that Corwin's "highly political travelogue" was strong in bringing "directly to the people of one country the voices of people in other lands"; however, it suffered from an "excess of personal narration." *Time* said that although the series sometimes could "lag, repeat itself, get incoherent," it boasted a "wonderfully perceptive, intimate sound track." *Newsweek* opined that Corwin at times sounded "labored and pretentious, cutting down the pace of what manages to remain one of the best bits of personal and exciting radio reporting yet heard." In *The Nation,* Lou Frankel observed that recorded actuality was "not enough to provide topnotch dramatic entertainment." In contrast to Corwin's earlier, finely-honed radio dramas, "Corwin's audience find themselves getting reality instead; and reality has many rough edges."[79]

Perhaps that was reflected in the audience response. CBS reported that it received 689 letters from across North America, the brunt of them laudatory. "Can't we have more such programs, and less of the moron-bait that befouls the airwaves?" wrote one correspondent. Another wrote to Corwin, "If the human race is to survive, we must have more people like you to wake America up." Still, the mail paled in comparison with the more than four thousand letters that CBS had received two years earlier in response to *On a Note of Triumph*.[80] In addition, whereas *One World Flight*'s premiere earned a 3.6 rating, its NBC competition, Bob Hope, earned a 30.6 rating. That still translated into more than two million listeners for Corwin, but as one critic noted at the time, it suggested that the American radio audience on the whole was far more interested in amusement than in hard questions about the world's future. Furthermore, the numbers did not substantially improve during the run of Corwin's series.[81]

CBS's scheduling of *One World Flight* against Hope sparked some consternation. Just before Corwin left on his trip, *Variety* reported that the *Time* publisher Henry Luce was negotiating via the Young and Rubicam advertising agency to sponsor Corwin's upcoming series and spend up to one hundred thousand dollars on it. A potential obstacle was Corwin's insistence "that he would have sole control of what goes on the air." Within a couple of weeks, the negotiations had fallen apart.[82] While Corwin was overseas, it was again reported that a "half dozen big bankrollers" were interested in sponsoring the series, once more with the stipulation that Corwin would have the final say over any such deal as well as a promise from CBS of "an absolutely free hand" in the series' content and format.[83]

As it happened, Corwin received no sponsorship for *One World Flight,* nor did he ever want it for any of his other programs: "I was offered it at five times what I was earning. And I had the good sense to decline, because that would have meant the abandonment of the freedoms that I enjoyed."[84] With such freedoms, though, came a downside—sustaining programs such as Corwin's tended to air at times when few people were likely to listen. In fact, on three previous occasions, Corwin's programs had been scheduled at the same time as Bob Hope's, who consistently had one of the highest-rated series on radio. That prompted Corwin to comment ruefully in an essay that "your writer seemed forever in the position of hoping against Hope."[85]

Even so, Corwin had reason to expect matters to be different with *One World Flight.* At the start of 1946, CBS had indicated that it would raise the status of sustaining programs.[86] Then in March, the Federal Communications Commission had issued its Blue Book report, outlining broadcasters' public-service responsibilities. As will be discussed in more detail in the next chapter, the Blue Book called on the networks and their affiliated stations to devote more time to nonsponsored shows. Broadcasters attacked the document and its authors as leftist, with the activist FCC commissioner Clifford Durr a particular target. Corwin defended Durr and government regulation of broadcasting in his Beverly Hills speech prior to his departure: "Your radio, like your community and your country, is largely free and decent, but it is not that way by natural law, like the force of gravity; it must be patrolled, just as police cars patrol the free and decent community of Beverly Hills."[87] William Paley had seemed to agree in a speech in October 1946 when he called on broadcasters to police themselves more rigorously and produce innovative new documentaries. According to *Variety,* the prevailing wisdom was that *One World Flight* embodied "the new type of programming so recently mapped by Paley himself" and therefore would air at "peak listening time."[88]

Airing against Bob Hope did not exactly constitute a "peak" slot. Years later, Corwin would say that he did not mind *One World Flight* being put opposite Hope because his audience and Hope's were "mutually exclusive."[89] At the time, though, he was "reportedly miffed," and the scheduling was a minor cause célèbre among liberals. The *New Republic,* edited by Henry Wallace, had reported that the networks were "dropping liberal commentators from coast to coast," and many saw CBS's scheduling of Corwin's series as reflecting that same trend.[90] In response to such criticism, some of it coming from within the network itself, an anonymous CBS executive said that they had given Corwin "carte blanche, invested lavishly in time, staff, and preparation expenses, [and broken] our own policy against recordings" to make "a first-class gesture to socially conscious broadcasting." Complaints about the scheduling were therefore "downright ungrateful." *Variety* summarized CBS's position as being that "no show, sustaining or commercial, can be guaranteed a major share of audience or protected against competition."[91]

Regardless, the relatively small listenership for *One World Flight* limited its impact. There were also limitations to the series itself. Even while still overseas, Corwin lamented in a letter to his family "the terrible superficiality" of a trip whose crowded and rushed itinerary prohibited him from fully grasping any one country's complexities. In a news conference following his return, he jokingly said that "few travelers have seen less" than he had and that he had missed such notable attractions as the Taj Mahal, the Veil of Kashmir, and Joseph Stalin. He later said that he wished he had asked better questions in his interviews.[92]

In some ways, he also may have been naive; as his friend Ed Murrow observed, even though Corwin was "a great dramatist and a fine writer," he was perhaps not "the best reporter that radio has ever had." Murrow had been fueled with the same One World idealism as Corwin immediately after the war, but he quickly grew disillusioned with the Russians, especially after Soviet-backed communists ousted the Benes government in Czechoslovakia in 1948.[93] "I was never a cheerleader for the Russian way of doing things," Corwin wrote at the time of *One World Flight* (while adding that he also "always had certain reservations about the way the British, Dutch, Swiss, and we Americans do things").[94] Later he would note the Russians' "xenophobia" and their repression of free political and artistic expression. Still, in *One World Flight* he viewed the Soviet Union and especially Czechoslovakia with optimism, reflecting the views of many liberals who admired how Russia had fought the Nazis and who hoped for a postwar rapprochement with the Soviets.[95]

Beyond that, the speed with which Corwin wrote *One World Flight* and his unfamiliarity with using recorded actualities resulted in a series that at times sounded less like a unified work than it did "a radio newsreel full of bits and pieces," as one critic described it.[96] His "murals and polemics" style tended to focus on abstractions instead of individuals. As another observer wrote, Corwin's sympathy was "not for the child under [his] hand but for nameless and countless children everywhere."[97] So it was that in *One World Flight*'s premiere, the weeping Italian woman (identified only as "the poor widow Camelia") was made to represent the "symbol of hopelessness." Such a style did not play as well as it might to radio's unique ability to tell personal, intimate stories.[98]

None of that diminishes what Corwin still was able to achieve. In effect, *One World Flight* created a new genre that was ahead of its time—the actuality-based long-form audio program that years later would find a home on American public radio. Corwin overcame daunting obstacles to bring sounds from around the globe to an American audience on an unprecedented scale. He provided a snapshot of the remarkable transformations underway in China, India, and the Philippines in addition to central and eastern Europe. Moreover, he freely expressed his sense that, regardless of his desire for peace, there still was "an awful lot of bubbling left in the cauldron."[99] Corwin's aim at fair-mindedness prompted him to include perspectives that directly contradicted his own; as Jerome Lawrence observed in *Hollywood Quarterly*, "He gives us the good and the bad, the democrat and the fascist, the builder and the opportunist."[100] He also made it clear that he was offering his thoughts based upon only brief visits to each country, as in the Russian program, in which he said that he preferred "to leave broad social conclusions to observers who have been there not weeks or months, but years."[101] In that, he was being more modest and transparent than some journalists, who might have made more authoritative pronouncements on the basis of similarly limited observations.

The rough-edged reality that emerged from *One World Flight* encouraged listeners to draw their own conclusions in a way that earlier, more didactic works, such as *On a Note of Triumph*, had not. One critic said that the series showed that "the natural dignity within man makes him capable of an understanding of the problems confronting mankind," whereas another said that it indicated that hopes for peace were "offset by the despair engendered by those who adhere to intolerance."[102] Such were the anxieties and ambiguities of a world just then drifting into the cold war, and Corwin gave them vivid voice in *One World Flight*. "It was good to send a poet around the world," wrote Jerome Lawrence in 1947. "He has a way of listening to the rhythms of tomorrow."[103]

* * *

For Corwin, life's rhythms grew more agitated following his return from his round-the-world trip. With *One World Flight*, he had taken advantage of a historical moment when key forces aligned: the brief carryover of Wallace- and Willkie-style liberalism, flush with the victory over fascism; the novelty of portable, if decidedly imperfect, audio recorders; the increased regulatory scrutiny that encouraged the networks to demonstrate social responsibility, if not necessarily when large audiences were listening; the last flourishing of sustaining network radio programming.

By the time the series concluded in the spring of 1947, that moment already was starting to pass. Doing what Corwin had sought to do—a "blunt and candid laying bare of inequities, shortcomings, and injustices"[104] combined with a denunciation of cold-war militarism and paranoia—would become exponentially more difficult. Soon after *One World Flight* had finished airing, the House Committee on Un-American Activities subpoenaed Corwin's scripts for the series. *Variety* reported that CBS "met the request deadpan, merely notifying its legal department to prepare the copies" of the scripts. Corwin later said that nothing came of the matter.[105]

Yet he was becoming more outspoken. In July 1947, he was keynote speaker at a conference on "Thought Control in the USA" sponsored by the Progressive Citizens of America, which the following year would back Henry Wallace for president. "If you stand for One World," Corwin said, "[or] if you believe literally what is said in the great documents of freedom upon which the United States and the United Nations are established, then you are suspect of participation in a colossal international Communist Front."[106] That same month, an anticommunist (and according to Corwin, profascist) magazine editor testified before HUAC that Corwin had communist sympathies. In October, Corwin cowrote and directed the radio program *Hollywood Fights Back* in response to HUAC's investigations of the movie industry. "I notice from time to time that your name is bandied about in the public prints as a suspicious, if not a dangerous character," CBS's Bill Costello wrote him that year from Japan. "These are bad times for people with open minds. . . . It's all very well to be able to see neo-fascism sprout like weeds all over the globe, but it's tragic when nobody wants to hear about it."[107]

By the end of 1947, Corwin was targeted by the red-baiting newsletter *Counterattack*. A report on Corwin in its files mocked his affinity for "the 'peepul' and the 'brotherhood of man'" by equating it to "the Soviet definition of the proletariat." Going to Moscow to meet with artists proved his "tender, unswerving devotion to the Kremlin," as did his criticisms of HUAC and the favorable coverage the communist newspaper the *Daily Worker* had given to *One World Flight*. Corwin's "record of participation in subversive activity" was "staggering not only in quantity, but even more significant—in quality"; his "propaganda prettily penned" had contributed significantly to "the disarmament of the American people."[108]

The increasingly reactionary political climate and growing financial pressures in the broadcasting industry would come to a head the following year for Corwin at CBS, representing the logical conclusion to what had been charged at the height of the *One World Flight* scheduling controversy—"that CBS, after years of 'showcasing' Corwin as its top concession to liberalism, now finds itself stuck with him—and wishes it wasn't."[109] The ambitious CBS plan to send other dramatists to report on the world's peoples would not come to pass. On the domestic front, though, liberalism would still find a vigorous voice at the network throughout 1947 via an ambitious endeavor bearing the imprimatur of not only Edward R. Murrow but also William Paley himself.

3

New and Sparkling Ideas

The creation of the CBS Documentary Unit in 1946 reflected contradictory impulses within CBS and its head, William Paley. A common historical assessment of Paley and the network in the immediate postwar years is that they largely abandoned the pretense of public service in favor of finally seizing the competitive edge over their archrival NBC.[1] Paley wrote in his memoirs that he rejected a proposal in 1945 by his second-in-command Paul Kesten to transform CBS from "a mass medium into an elite network" that would reject crass commercialism in favor of quality—in effect, the kind of programming represented by Norman Corwin. To Paley's mind, such a plan would result in a "narrower, specialized network of dubious potential," whereas he "was determined that CBS would overtake NBC as the number one radio network."[2] Soon after, Kesten was replaced by Frank Stanton, and CBS began luring high-rated entertainers and shows away from NBC.

At the same time, Paley "believed that he could have it all," as one historian has written; he could at once "surpass his commercial rivals *and* hold his head high among his more cerebral friends."[3] There was also residual idealism from the war, during which Paley had served as a lieutenant colonel in Europe while socializing with Edward R. Murrow and other CBS journalists and witnessing the horrors of the Dachau extermination camp just after its liberation. In the words of CBS's Robert Lewis Shayon, for a time afterward, "the smell of wartime clung to the men in the grey flannel suits, and in their civvies, they continued to honor the national purpose."[4] So it was that Paley at the end of 1945 promoted Murrow to network vice president of public affairs in what was seen at the time as "the first major attempt to elevate [noncommercially sponsored] sustaining fare to the same high level of commercial programs" and with a comparable level of funding."[5]

The Blue Book and the Documentary Unit

Paley and CBS's efforts in that direction would be motivated by more than altruism. According to *New York Times* radio critic Jack Gould, 1946 would see radio "subjected to more diverse and insistent criticism than the industry experienced in the whole of its previous twenty-five years, the main burden of the complaint against the ethereal art being excessive commercialism."[6] Among other things, Gould noted the popularity of the novel *The Hucksters,* which satirized the influence of advertising on radio. Many others weighed in on broadcasting's shortcomings. The New York University professor Charles Siepmann, who had formerly worked for the BBC, wrote *Radio's Second Chance,* in which he charged that the American networks had "largely abdicated to the interests and point of view of [advertising] agencies and firms that have become more masters than clients." Meanwhile, sustaining programs had grown "subject more and more to the slings and arrows of a most outrageous and paradoxical fortune," such as being aired at times when few were likely to listen.[7]

Siepmann was a consultant on the most controversial critique of radio in 1946, the so-called Blue Book issued in March by the Federal Communications Commission. The report quoted Frank Stanton's testimony before Congress that CBS's sustaining programming represented its "greatest contributions to network radio broadcasting" and helped it achieve "a full and balanced network service." At the same time, the FCC report said that many local stations opted not to broadcast such network offerings and instead aired more commercially lucrative programs. The report also graphically described radio's rampant commercialism, noting that the industry had drifted far from the purported standards of 1930, when William Paley had told Congress that CBS prohibited "overloading of a program with advertising matter." The Blue Book indicated that future broadcast-license renewals would rest in large part on the extent to which stations curbed "advertising excesses" and aired sustaining programs dealing with substantive public issues.[8]

Some praised the Blue Book when it first appeared. "Broadcasters now must face the fact that radio cannot operate under the same set of rules as those which govern other business operations," editorialized *Variety,* adding that up until then, radio had "only paid lip service to the responsibility inherent in its use of" a public commodity and that the report might "be a blessing in disguise" in revitalizing the medium.⁹ The National Association of Broadcasters (NAB), though, launched a sustained, vitriolic assault on the report and those who had contributed to it, including Siepmann and the FCC commissioner Clifford Durr. The NAB's president Justin Miller warned that through the Blue Book, "the way will have been paved for the government to take radio over—and to take the press and motion pictures over. And that, I submit, is the Communist technique."[10]

William Paley addressed the Blue Book debate in a speech to the NAB annual convention that October. On the one hand, he declared his "fundamental opposition" to the notion "that a Government agency should have the power to blueprint the kind of radio program which the American people shall hear," saying that "a medium which gives most of the people what they want most of the time" was being unfairly maligned. That was in keeping with past CBS battles against increased regulation. On the other hand, Paley acknowledged that the Blue Book had been correct in charging that radio had engaged in "advertising excesses" that were "irritating, offensive, or in bad taste," and he called on broadcasters to police themselves more stringently: "Our real task is to earn and hold public confidence by deserving it, matching with our own responsibility the responsibility we ask of critics." He added that broadcasters should invest more time and energy into public-service programming: "[N]ew and sparkling ideas in the presentation of educational, documentary, and controversial issues is one of our greatest challenges today." Consciously or not, Paley's call for "new and sparkling ideas" echoed Paul Kesten's proposal from the previous year to elevate CBS's programming.[11]

It also aligned neatly with a major new endeavor that CBS had announced the month before the NAB convention. A new Documentary Unit that CBS publicity said was "devoted exclusively to the production of programs dealing with major domestic and international issues and involving extraordinary research and preparation" was established. Its goal was to produce ten to twelve broadcasts each year on subjects potentially ranging "from atomic energy to the Nuremberg Trials, the housing crisis, juvenile delinquency, occupation policies in conquered nations, or any other problems or issues which deserve widespread public attention." Just as also would be the case with Norman Corwin's *One World Flight*, the Documentary Unit was seen as implementing the proposals that Paley put forth in his NAB address.[12] In addition, it reflected the increased emphasis on sustaining programs under Murrow's executive oversight. One of his pet projects would be a new program critically examining the print news media, *CBS Views the Press*, hosted by Don Hollenbeck.[13] The Documentary Unit would be another such project.

Murrow called it "an involved, difficult, expensive, and altogether obvious thing to do"—obvious in that it logically built upon CBS's war coverage in addition to representing the kind of journalism that the quality print media had long done.[14]

CBS did a trial broadcast of sorts in October 1946 with *The Empty Noose*, which was timed to air the night of the hanging of convicted war criminals at Nuremberg. The author was Arnold Perl, a self-identified longtime "antifascist" who said that although he was frustrated that "every minor blow sounded for decency and progress on the air has been like pulling teeth," he nonetheless was committed "to stay in radio and try to do something about it."[15] *The Empty Noose* used no recordings of real-life participants in the Nuremberg Trials. Instead, the broadcast featured an actor playing an American G.I. "eyewitness" to the hangings. He recounted in excruciating detail the atrocities that the Nazi war criminals had perpetrated, from turning a Greek Orthodox church into a mass urinal to using the skin of Holocaust victims for lampshades. In the program's climax, the eyewitness spoke of "an *empty* noose . . . waiting to choke off the last breath of the foulest thing we'll ever know—Fascism." He said that the potential seeds of fascism in America lurked in the red-baiting of union members, hate crimes against nonwhites and non-Protestants, and renewed calls for war: "Tonight at Nuremberg—and tomorrow—there will still be one round coil of rope ready to be used. It's going to take a lot of self-examining, a lot of faith in what we believe in, a lot of willingness to fight for it, a lot of speaking out, for all of us, here and everywhere, before that empty noose is filled, and we can say we have won, we have conquered. [Pause.] I think we can do it."[16]

A reviewer praised *The Empty Noose* for "combining the best in radio with the most enlightened in citizenship," adding, "It's as if CBS' returnees [from the war] had stood up to remind themselves, and the net's listeners, that maybe the war's lessons are of a lasting nature."[17] Perhaps more significant to CBS management was the audience response. The network preempted the *Ellery Queen* detective show to air *The Empty Noose* on short notice, to the displeasure of many CBS affiliates. Still, the ratings suggested that the program reached nearly six million listeners, only a small decline from the typical *Ellery Queen* audience. According to *Variety*, CBS was thus convinced that it could air future sustaining programs "in similar choice middle-of-the-evening segments, cancelling out braces of half-hour sponsored shows that have maximum audience pull; but never two in the same slot, and none on a regular schedule"—instead, only when the programs were ready to air.[18]

With that plan in place, the Documentary Unit went to work. Although the unit would collaborate with a number of outside writers and producers, Murrow appointed four CBS staffers to it full-time: Robert Lewis Shayon, Ruth Ashton, Lane Blackwell, and Robert Heller. Ashton later described the group as "spoiled folks," adding, "We were hated, literally, by some of the [other] writers because we were given the best assignments and the best [air] times and the best treatment and maybe the best money."[19] Heller served as the unit's director. He was a thirty-one-year-old Harvard graduate who had worked with Norman Corwin on the *This Is War!* radio series and with the movie director Frank Capra on the *Why We Fight* propaganda films. He also had produced other war-themed programs for CBS, including *The Empty Noose*.

Heller would become the unit's most fervent booster. He credited Paley and Murrow with helping to launch "a profound revolution in the broadcasting industry's attitudes and techniques" by creating "a virtual Utopia for craftsmen who believe in radio's usefulness as a social force." Never before, he wrote, had a network created a wholly autonomous group free of deadlines and at liberty to experiment while being supported with the resources for extensive travel and investigation. The result was "the birth of a new kind of journalism, more ambitious, comprehensive, and vital than any other effort in print or sound."[20]

Robert Lewis Shayon and *The Eagle's Brood*

Shayon was assigned to research and write the unit's debut production. He was thirty-four, having joined CBS in 1942 to work on series promoting the war effort before traveling to Europe and meeting Murrow. Shayon returned home newly determined to see the United States and radio uphold what he saw as their responsibilities. He told of the misery he had just seen overseas in a 1945 script for the radio series *The Land Is Bright*. "My country, you're young and you're brave and you've got power and wealth," wrote Shayon. "But the world is closing in on you fast. We must roll up our sleeves and dig into our hearts and our pockets and work with the men and women of the world to clean out the swamps of the world." At about the same time, he publicly chided the radio industry in a piece for *Variety*. "Keep on selling soap, boys— that's right and proper—but for God's sake give a little thought and network time to selling peace," he wrote. "Those youngsters we saw on the rubble heaps of Europe and your own kids in your backyard will thank you."[21]

The following year, Murrow freed Shayon of all other responsibilities for several months so that Shayon could create *Operation Crossroads*. The hour-long program aired in May 1946, not long before the Bikini atomic-bomb tests. It featured a cast including Henry Wallace, Harold Ickes, Harold Stassen, Supreme Court Justice William O. Douglas, the atomic physicist Harold Urey, and Albert Einstein, whom Shayon personally recruited during a visit to Princeton. The cast engaged in scripted dialogue with a group of ordinary citizens specially assembled for the program, addressing common questions and fears about the atomic bomb and calling for international control of the weapon. "We must remember that if the animal part of human nature is our foe, the thinking part is our friend," said Einstein. "We can and must use [our ability to reason] now, or human society will disappear in a new and terrible dark age of mankind—perhaps forever." Archibald MacLeish concluded the program by asserting that the people's collective will should be determined not "by reading the papers" or listening to "an advertising slogan" but "by talking and listening and thinking." That way, Americans could best learn how to live "in a new and greatly threatening, greatly promising world."[22]

The success of *Operation Crossroads* helped convince CBS to create the Documentary Unit and put Shayon to work on its first program, *The Eagle's Brood*.[23] It would tackle a subject of controversy in radio and the country at large. One of the many criticisms aimed at radio right after the war was that it glamorized crime and contributed to juvenile delinquency. In one example, a judge blamed crime shows for influencing the teenaged William Heirens, charged in 1946 with being Chicago's notorious "Lipstick Killer."[24] In response to such concerns, broadcasters adopted a new code and added more socially conscious storylines to programs such as *Superman*. William Paley discussed the controversy in his NAB speech, saying that the solution was not to ban crime programs but rather to improve their writing.[25]

The Eagle's Brood would be another radio response to worries about delinquency. The directive to tackle the subject came from Murrow rather than Paley and reflected the fact that the worries extended well beyond radio's influence on youth.[26] President Truman at the start of 1946 had condemned a "deplorable" rise in juvenile crime and called on his attorney general Tom Clark to help generate solutions. Clark convened a national conference to discuss the issue, having declared that the country potentially faced "a wave of delinquency such as never before experienced in its history."[27] The press amplified such alarms. *Life* published a ten-page photo essay showing youths loitering, vandalizing, fighting, stealing, and landing in jail. As for a solution, the magazine asserted that "prevention rests with the parents."[28]

In *The Eagle's Brood*, Shayon would adopt a broader viewpoint that was heavily influenced by G. Howland Shaw, the chair of an antidelinquency committee formed out of Attorney General Clark's national conference.

Shaw was also on the board of the Industrial Areas Foundation (IAF) of Chicago. Rather than primarily blaming parents for delinquency or relying on child experts to dictate solutions, the IAF advocated the neighborhood-council approach of Saul Alinsky.[29] He had cofounded the Back of the Yards Neighborhood Council in one of Chicago's most crime-ridden districts, with the IAF extending the council's work. Alinsky argued that to fight juvenile crime, one must also fight "the crime of economic insecurity; the crime of poor housing; the crime of inadequate medical care; the crime of prejudice and man's inhumanity to man."[30] That fight had to happen at the grassroots, with community members directing "all of their efforts and collective skill towards the solution of their common problems." To that end, Alinsky called for additional "People's Organizations" like the Back of the Yards Council, which he boasted had "bridged all of the economic, social, religious, and political cleavages" in the neighborhood while promoting "the responsibility, strength, and human dignity which constitute the heritage of free citizens of a democracy."[31]

Alinsky proclaimed himself a radical in his bestselling book *Reveille for Radicals*. He dismissed liberals in comparison as those who "talk glibly of a people lifting themselves by their own bootstraps but fail to realize that nothing can be lifted or moved except through power." Such talk worried some on the left, who feared that it could lend itself to demagoguery or even fascism.[32] Others found Alinsky reassuring. He had long courted the press, and for better or worse, the resulting coverage tended to emphasize "people lifting themselves." The solidly Republican *New York Herald Tribune* commended the IAF for promoting "uncoerced self-improvement . . . that no government paternalism could supply." Similarly admiring stories appeared in *Reader's Digest* and *Woman's Home Companion*. In a six-part *Washington Post* series, Agnes Meyer praised the IAF for promoting "the most powerful upsurge of organized individualism yet to come into being in the USA."[33] Regardless, according to his biographer, Alinsky's philosophy was consistent with a "left liberal, New Deal agenda"; it stressed that "one questioned authority, took the initiative to address community problems," and understood "how events and forces in the larger world affected one's own life and community."[34]

Howland Shaw literally guided Shayon toward the same philosophy in the making of *The Eagle's Brood*. He arranged a nationwide fact-finding trip for Shayon in the fall of 1946 with Edward R. Murrow's blessing. Murrow later remarked that CBS could have created a perfectly adequate juvenile-delinquency show without sending Shayon on the road, but the trip gave the program added authenticity: "It was, as journalists sometimes say, the nail holes in the wrist."[35] Shayon visited New York, New Orleans, San Francisco, Los Angeles, Denver, Boston, and Washington, D.C.; he also toured Alcatraz, a death-row cell in the South, and other correctional facilities, where he learned of children routinely being flogged.[36] "What I saw and heard cast me deeper and deeper into despair of ever finding a way out of the abysmal social mess," he recalled. That changed when he arrived at the final stop that Shaw had arranged for him: a visit with Saul Alinsky in Chicago's Back of the Yards neighborhood. "I lit up like an electric tree," said Shayon, "because in his neighborhood council I saw a vision of hope, of democracy solving its own problems."[37]

Shayon returned to New York determined to "attack the public indifference to the whole matter" via a program intended "to stir the listener, to unsettle him, to shock, to enlighten, and finally to inspire."[38] *The Eagle's Brood* (originally titled *Citizen's Journey*) would follow a format similar to his recounting of his wartime travels in *The Land Is Bright*. Although Shayon later said that he had taken "careful, copious notes" while touring America, he had not followed the lead of Norman Corwin, who had struggled so mightily with a wire recorder while gathering material for *One World Flight*. Instead, as with *The Land Is Bright, The Empty Noose,* and most other such programs of the day, *The Eagles' Brood* took the form of docudrama, both because of the difficulties with field-recording technology and the recording ban that was still largely intact at CBS and NBC. "Reality was simulated and transformed," said Shayon, with actors, music, and sound effects. That had an upside. As one historian describes it, it encouraged "innovative artistry" among radio writers who sought to "convey socially charged messages" in storylines that "leapt across space and time."[39]

"What a privilege for a documentary-maker," recalled Shayon of creating *The Eagle's Brood*; he had all the time and funding he needed.[40] That is not to say that he was wholly free of corporate oversight. In a December 1946 memo to William Paley, Murrow said that a few pages of the script had been rewritten to eliminate any "specific reference to press, radio, advertising, and other stimuli," thus avoiding implicating the media in fostering delinquency. CBS lawyers similarly directed Shayon to delete comments critical toward soap operas (some of which the network aired) and not to mention the names of the cities he had visited.[41] When CBS recruited the actor Joseph Cotten to play the lead role, the producer David O. Selznick excused him from filming a movie on the condition that there would be nothing "attributing juvenile delinquency even inferentially to the effect of motion pictures." At the same time, Selznick acknowledged "the great [publicity] value to me to a gesture on our part towards a juvenile delinquent program."[42]

CBS also seized upon the program's publicity potential. In the weeks leading up to the broadcast, the network issued news releases trumpeting that *The Eagle's Brood* would present a "picture of civic indifference, official impotence, and economic short-sightedness that is destined to shock the average hearer by its straightforward statement of authenticated facts."[43] The network president Frank Stanton wired the mayors of every city with a CBS-affiliated radio station and asked them to listen, and Murrow personally recorded promotional spots to air on the stations. CBS also announced that the General Federation of Women's Clubs had "called on all its members in 166 cities" to "arrange town meetings and forums to listen to the broadcast and to discuss, immediately afterward, ways of implementing the program's conclusions."[44]

After Shayon recruited the actor Luther Adler at the last moment to play a character modeled on Saul Alinsky, the live broadcast of *The Eagles' Brood* took place on March 5, 1947.[45] It was preceded by an announcement that *Information Please,* a commercially sponsored quiz show, would not be heard that evening. Then dramatic music was interspersed with the reenactment of a crime—"three husky American youngsters on the prowl" who were beating and robbing a drunk on a city waterfront. "From Boston to Butte, Hartford to Houston, Savannah to Spokane, more and more children are climbing the crescendo of a wave of crime, brutality, violence, and murder," an announcer intoned. "Every terrifying episode bears a direct, challenging meaning to you and yours. . . . This is what is happening to *all* of us—and to our children—today."[46]

Joseph Cotten then assumed Shayon's role and retraced his nationwide journey, with frequent and at times intentionally jarring shifts in time and location. First came a youth canteen in "the Negro quarter in an East Coast city" at Halloween. A fifteen-year-old gang member related how he already had survived two shootings and how the local gangs arranged "rumbles." At heart, though, the youths were "just like any other American youngsters" in "wanting normal fun," according to a recreation worker. Next there was a visit to another eastern city and a twenty-year-old mother who had deliberately burned her three-year-old son with an iron. The local press had made no effort to probe the woman's unhappy background and honed in only on the most sensational details of the case. "A good witch hunt now and then is like a bloodletting," an official said. "If you ask me—it's a good deal."

Cotten's narrator then recounted a grim series of experiences across the

South: a juvenile-court judge who "plays God with children's lives for votes," a prison farm of "unspeakably filthy camps and dungeons" where youths mixed with "murderers, thieves, dope addicts, [and] perverts," and worst of all, a death row with four teenagers awaiting execution. Over the voice of a deputy describing how the county was investing in "the most modern improvements" to its death chamber, Cotten's narrator expressed his thoughts with mounting fury: "It didn't matter that this was America. It didn't matter that this was the twentieth century. . . . Nothing mattered except this man's blind, stupid, blasphemous, incredible pride in his plaster paint and portable electric chair."

It was little better out West. A truant officer told of children falling through the cracks due to inadequate schools and unfair public-housing codes. A visit to the penitentiary at "the Rock" revealed that some of the nation's worst criminals were barely out of their teens. A reformatory superintendent related how youths were brutally beaten by prisoners and guards in local jails. Cotten's narrator ticked off a host of maladies: "ignorant, careless, or incapable parents—rich *or* poor," "lack of jobs," "indifferent citizens," materialistic social values. There seemed to be no solution: "It's a merry-go-round. Where do we get off?"

Finally, the narrator arrived in a "city in the Midwest," where he "met a man who answered my question." That was Luther Adler's Alinsky-inspired character, who declared that to make change happen, "*we're* going to have to do it with our *own* hands, brains, money, and imagination." A merchant, Mexican immigrant, union organizer, teacher, police captain, and war veteran took turns praising the local neighborhood council. It represented "the most exciting thing that's happening in America today," said Adler. "That's why we've got to have Neighborhood Councils everywhere in America." Through such councils, not only could citizens attack the root causes of delinquency and other ills, but they also could recognize their common lot and mission. "Our children are our only REAL assets," Adler concluded, as music swelled underneath him. "Our FUTURE is 'THE EAGLE'S BROOD.'"

Shayon recalled the immediate impact of *The Eagle's Brood* as being "theatrical," with CBS's switchboard flooded with calls.[47] Edward R. Murrow said that the network received "in the vicinity of 3,500 to 4,000 letters," along with reports of several neighborhood councils being formed; one community raised thirty thousand dollars in a week for a new youth center.[48] Alinsky wrote a friend that it was "as a fine and as outstanding a public acknowledgement of our work and philosophy as we can ever expect to get" and said that "CBS was being deluged with requests for identification" of the real-life model for Luther Adler's character in the program (Shayon ended up referring all such requests to Alinsky himself).[49]

The *New York Times* critic Jack Gould proclaimed of *The Eagle's Brood* that "the art of broadcasting found its voice and lifted it as one truly come of age," adding that the program showed how Americans "could build a backfire of common sense and common action."[50] *Time* commended CBS for preempting commercial shows to air the documentary and said that the network "had demonstrated that when radio has something to say about an important problem—and says it intelligently—people will listen."[51] Others similarly praised CBS for employing "radio's supreme capabilities for dramaturgy, sound effects, and musical compulsion" in making important issues "vital and significant to an uninformed public,"[52] and for demonstrating public-spiritedness on behalf of a "radio industry [that] has not been notable for providing such leadership."[53] The praise was not unqualified. Some who otherwise liked the program demurred that it suffered from "an overabundance of narration,"[54] that it "oversentimentalized" delinquency,[55] and that its neighborhood-councils agenda alone would not eliminate the problem.[56] One Boston critic noted that "to realize the utopia described in 'The Eagle's Brood' would take more money than citizens are now spending." The liberal newspaper *PM* went further in scolding the documentary for not delving more into the systemic inequities underlying delinquency.[57]

In addition, CBS's scheduling of the program raised some eyebrows, just as it had with *One World Flight* a couple of months previously. *The Eagle's Brood* achieved a 6.4 audience rating, which compared favorably with the ratings that CBS's commercial shows had received in the same time period and especially with the numbers that past documentaries had earned. However, the program aired against Bing Crosby's popular show, similar to the scheduling of *One World Flight* against Bob Hope.[58] One critic wrote that "all the people who should have heard 'The Eagle's Brood'" were instead listening to Crosby: "I'd have liked to have heard him myself."[59]

As it had with *One World Flight*, CBS sought to refute such criticisms. "Any program is in a tough spot if it faces competition," said Murrow, noting that *The Eagle's Brood* still had aired in prime time. CBS also touted the fact that 42 percent of those who had tuned in did not ordinarily listen to the network's programs during that time slot.[60] Regardless of how many people listened, the program appeared to have achieved one of CBS's key goals. A magazine executive wrote Frank Stanton that *The Eagle's Brood* had done "more for radio to win over its most severe critics than almost anything I can remember in the last decade."[61]

Ruth Ashton and *The Sunny Side of the Atom*

The Documentary Unit followed *The Eagles' Brood* with *A Long Life and a Merry One*, which aired in April 1947. Howard Rodman wrote it following a nationwide tour similar to Shayon's. According to network publicity, the documentary's focus was the "appalling contrast between the swift advances in medical science and the inadequacy of medical care for the American people." *Variety* noted that the program highlighted a multitude of problems, including the fact that "proper knowledge of preventive medical care and treatment is beyond the reach of 80 percent of the people in the country." Like its predecessor, *A Long Life and a Merry One* employed dramatizations, with Van Heflin playing the lead role. *Variety* praised CBS for having "injected a maximum of showmanship."[62]

The unit produced two additional programs in June. *Experiment in Living* reported on experiments at the University of Iowa and the Massachusetts Institute of Technology. As CBS described it, groups of ten-year-old boys were "exposed successively to the practices of autocracy, laissez-faire, and democracy." The autocratic groups "developed no community spirit, and were found to harbor deep resentments"; the laissez-faire ones "disintegrated into chaos and confusion"; and the democratic ones were "the happiest and most productive." A CBS consulting psychologist said that the program showed that democracy "runs the danger of losing its vitality if people take it too much for granted."[63]

The second June program returned to the docudrama model established by *The Eagle's Brood* and *A Long Life and a Merry One,* and it featured the work of the sole woman in the Documentary Unit. Ruth Ashton had joined CBS News before even graduating from Columbia University's journalism program. The CBS news director Paul White, who doubled as an instructor at Columbia, recruited her. After a stint writing news for Robert Trout, Ashton was assigned to produce a program called *Feature Story,* for which she did everything but actually read the stories on the air. According to her, CBS management felt that women's voices were too "squeaky" to be airworthy. (Years later, she would avow that she eventually was allowed to audition for an on-air position after it was discovered that alcohol lowered the pitch of her voice. However, on the day of her audition, the beer she drank turned out to be flat and failed to produce the desired effect; hence she remained off the air.[64])

It was not only that women did not read newscasts at CBS; there were comparatively few women in the entire news division, although Paul White had hired some in addition to Ashton as newswriters and desk assistants.[65] Regardless, Murrow tapped Ashton for the Documentary Unit when she was only twenty-four years old. As she recalled it, he took aside the unit members and "asked us each to choose the story closest to our heart [and] spend as much time as we needed to go do that story."[66] Ashton had excelled in math and science in school, and she had been deeply affected by the Hiroshima bombing, which she had written about when the initial bulletins came over the wire in the CBS newsroom. She asked to write about atomic science. Murrow was friends with David Lilienthal, the new head of the U.S. Atomic Energy Commission. According to Murrow at the time, "several of us [including presumably Lilienthal] became fed up with hearing dire tales of what would happen when the Empire State Building was toppled, or when the first atom bomb was dropped on Washington"—in effect, with the likes of *Unhappy Birthday* and similar recent radio programs that had issued dire warnings about a potential atomic holocaust.[67] Ashton was thus assigned to research a program called *The Sunny Side of the Atom* that, as she remembered Murrow describe it, would help listeners "unfreeze their minds by giving them information that atomic science is not all atomic bombs."[68]

Ashton was keen to gain the assistance of Albert Einstein, but her calls and letters to his office at Princeton went unanswered. Nothing if not determined, she finally went to Princeton herself. By chance, her cab driver spotted Einstein walking on the street, and Ashton got out to introduce herself. "Ah! The broadcasting lady!" Einstein replied. "He read the letters, apparently," recalled Ashton, and he graciously walked and chatted with her for several minutes, "but he didn't feel that a radio program could have any effect" in teaching people what was at stake concerning atomic energy; only face-to-face conversation could make a difference. It was not clear what, if anything, had changed in Einstein's mind following his participation the previous year in *Operation Crossroads,* but he at least had given her some insights that she could incorporate into the program.[69]

Ashton was able to interview many other central figures in atomic research and tour key facilities through the help of Murrow and his government contacts. During a trip lasting more than a month, she visited the Argonne Laboratory near Chicago and was among the first journalists permitted inside the Oak Ridge Laboratory in Tennessee. She also met with General Leslie Groves of the Manhattan Project that had developed the atomic bomb, and she interviewed Robert Oppenheimer (whom she recalled as "a poetic figure") at his office at the University of California in Berkeley. Ashton learned of cutting-edge research into medical advances that drew upon atomic science.[70]

When she returned to New York and CBS, tensions began to develop with the head of the Documentary Unit, Robert Heller. She recalled that he "felt I had secrets of the world in my head" and "almost put me in seclusion," fearing that if she talked to any outsiders about what she had seen or heard, "NBC would do something" with the story before CBS did. Even after she fell ill, she was told at first that she could not take time off from work until she had reported back fully all that she had learned. Ashton then began to collaborate with the writer-director Carl Beier on a script, but again there were tensions. As she remembered, Heller proposed doing the story "as kind of a fantasy, you know, Alice in Wonderland goes tromping" giddily through the new atomic age. An incensed Ashton found the idea "practically irreligious" and appealed to Murrow, who sided with her. The program instead took the form of a more straightforward dramatization of Ashton's trip and findings, similar to what had been done with *The Eagle's Brood*.[71]

The Sunny Side of the Atom aired on June 30, 1947. For once, CBS broadcast a documentary in an advantageous slot, preempting the *Lux Radio Theater*, which was one of its highest-rated shows with an average weekly audience of thirty-two million listeners.[72] The network again recruited a high-profile actor to play the lead role. "This is Agnes Moorehead, bringing you the most important story of our time," she said by way of introduction. "Mention atomic energy and you think of the atomic bomb and destruction. But there is another side to the atom." The program then re-created Ashton's travels. She followed a lead-bottle shipment from Oak Ridge to a doctor's office in New York City, where the shipment of radioactive tracers allowed doctors to measure the blood circulation in a "golf pro at the club back home" who faced the potential loss of his foot. The tracers revealed his circulation to be good enough that amputation would be unnecessary. "Thanks, Doc. I feel pretty good for a guy who's had atoms running around in his bloodstream," the golf pro said afterward.[73]

Next, the documentary recounted Ashton's trip to Berkeley and her conversation with Dr. Hardin Jones (unlike *The Eagle's Brood*, *The Sunny Side of the Atom* used people's actual names, though all were played by actors). Jones cautioned that "we oughtn't to pin too many hopes of miraculous cures" on atomic science. Still, it held enormous potential—"there's even a possibility that we may be able to create a living cell!" Elsewhere on campus, Ashton donned protective clothing to tour a bacteriology laboratory. "We're trying to disarm many diseases. But the world can commit suicide with them first," a researcher told her. "I think we may be able to make people want to live—in a world without disease and fear."

That was followed by a visit to an oil field out "where the South meets the Southwest." Geologists were using neutron instruments to measure the possible presence of oil underground. Their successes in detecting additional oil in older wells cheered one longtime resident of the area. "I can just hear them [derricks] all whumpin' away like they used to when they was young," he told Ashton. "Guess folks around here won't feel so bad about that atom bomb when they hear about this!" Next came a visit to an "atomic farm" where research was taking place to try to "speed the growth of food enough to feed every hungry mouth in the world," as a researcher there said.

Indeed, according to Ashton's sources, the atomic age offered a world of promise: it potentially could "propel rockets into planetary space"; "irrigate the whole of Arabia" and thus "solve the Palestine problem"; help finally give China and India "living conditions suitable for human beings"; allow farmers "to work a lot less, and produce a lot more"; and lead the way "to curing diseases now considered incurable." In Robert Oppenheimer's words, the possibilities that atomic science presented could "change the face of our world. But now we must work for them."

Ashton fell into a reverie as CBS studio musicians played underneath her "warmly and dreamily," according to the music cue in the script. She had a vision of "men standing straight and tall and confident" who were "urged forward by a new hope" and "for whom hunger was history, disease a word in the dictionary, war an abandoned answer to human problems." Then the music shifted "suddenly and ominously" as she remembered other things her sources had told her—that "the fuel that could power a city could be made into bombs to cripple a nation overnight," that radiation could "create abnormalities in future generations that will weaken the human race," that tens of thousands of atomic bomb survivors were already suffering the aftereffects: "You saw atomic science attacking cancer. We may have created it, in Japan."

The Sunny Side of the Atom concluded by briefly recapping Ashton's talks with David Lilienthal and Albert Einstein. Lilienthal helped convince her that "we must keep aware of what is being done with the atom in our government, and between governments, because if we don't, great decisions that must be made will be made by a handful of men behind closed doors." Ashton then recalled Einstein's "very words: 'Only talk between men promotes feeling in the heart.' When we are clear in heart and mind—only then shall we find courage to surmount the fear which haunts the world." Her journey had instilled in her an overriding faith: "We are bigger than the atom, and if we face the future boldly, we will enter a world made bright by the sunny side of the atom."

The program received mixed notices. *Variety*, which had praised the Documentary Unit's previous productions, panned this one as being "pretentious," "pedestrian," and "dull." It called Moorehead's performance "appealing and capable" but said that "a feminine voice immediately couched the show on a too-simple plane for the serious (though 'sunny') story at hand."[74] Ashton ruefully said later that it had been a mistake for her to meet with Moorehead before the program aired, "because I was quite young, and she then didn't sound as authoritative [in her on-air performance] as they would have liked to have had somebody sound, because she was being Ruth." That inadvertently made the subject "sound less serious and significant than it was," which was what Ashton had tried so hard to avoid.[75]

Other critics were more sympathetic. The *New York Times* said that the program deserved "prolonged applause for its intelligent and timely exposition." *Newsweek* wrote that it represented "the first time that the fundamental peacetime promises of atomic power had been adequately aired." It allowed that "Miss Moorehead's speeches were often long and occasionally tedious" but noted that "entertainment was not the chief purpose of the program." That was underscored by the reviewer for the *Christian Century*, who praised the show for overcoming "the paralyzing mediocrity" of most other radio offerings, adding that "in both conception and execution it met the sternest tests of social usefulness. Why can't [radio] do that sort of thing more often? There is, we are sure, a great public which would welcome it."[76]

Radio Research: Ernest Dichter and "Big Annie"

Such comments must have been precisely what Robert Heller wanted to hear, for he believed that radio documentary should be devoted to "extending the horizons of constructive citizenship." He asserted that in order to make certain that it was accomplishing that mission, researchers should test every documentary to determine how well it was holding its audiences and eliciting the desired responses. "Without this constant check of listener reaction," Heller wrote, there would be "no blueprint for improvement" in radio documentary's effectiveness.[77]

In fact, the Documentary Unit had been formed during what the *New Yorker* called a "frenzy" surrounding radio research: "All the networks are currently measuring, polling, quizzing, surveying, studying, testing, sounding out, analyzing, and otherwise probing the public in an effort to find out where radio stands."[78] Under president Frank Stanton, CBS was a leader in that research. Robert Lewis Shayon would work with the motivation researcher Ernest Dichter and with "Big Annie," the "Program Analyzer" invented by Stanton and Paul Lazarsfeld.

In time, Ernest Dichter would be hailed and vilified as a profoundly influential pioneer in the use of psychoanalytic techniques in marketing. Dichter himself termed it "the therapeutic job of democratic engineering," dedicated to moving people "out of the false paradise of knowledgeless animal happiness into the real paradise of the life of change and progress."[79] Translated, that meant tapping into consumers' unconscious desires in order to sell everything from soap and cars to cigarette holders, Jello, and prunes. It prompted Vance Packard in the 1950s to brand Dichter a leader among a new, dangerous breed of "hidden persuaders."[80]

When Shayon met him in 1946, though, Dichter was still a little-known CBS researcher from Vienna who had been studying the latent psychological values in the network's children's serials. Most of CBS's writers and producers tried to avoid him. Shayon nonetheless was "intrigued by his confidence and air of authority," and he hired Dichter to conduct interviews with ordinary people during the preparation of *Operation Crossroads*. The findings were "that the American mind was schizophrenic on the subject of the atom bomb," fearful of its destructive force but fascinated by atomic energy's potential. Shayon thus designed the program as "a national therapy session on radio."[81] In Dichter's words, it aimed to bring the listener "face to face with the inescapable facts of reality and offer him not only the tools with which to solve his problem, but also the encouragement to solve it." Dichter said that such an approach raised radio above the "medicine-man level" by using the tools of psychoanalysis and advertising in creating and promoting socially conscious programming.[82]

As Shayon began work on *The Eagle's Brood*, he again turned to Dichter to do interviews about delinquency. In a report on his findings, Dichter recommended that the program encourage listeners to accept personal responsibility for the problem. He suggested that no "blames of any kind, no matter how justifiable, should be pinned on anyone" and that no "remedies in the forms of clubs and agencies, etc., no matter how helpful and realistic, should be suggested as final means." Instead, the program should impress upon the listener that the "problem of juvenile delinquency is a close and personal one. It is a proof of the wrong reality of your own life. It is unreasonable and unrealistic to assume that you are not your brother's keeper."

As it turned out, Shayon decided against using Dichter's "therapy" approach for *The Eagle's Brood,* believing it "too abstract" and "not dramatic enough." Its "no remedies" recommendation also was not consistent with Shayon's advocacy of neighborhood councils.[83]

Once *The Eagle's Brood* aired, Shayon again came into contact with CBS's research department in the form of the Lazarsfeld-Stanton Program Analyzer. Stanton had come to CBS as a researcher himself in the mid-1930s and quickly rose through the corporate ranks. He became acquainted with Lazarsfeld through the latter's association with what would become the Bureau of Applied Social Research at Columbia University. Lazarsfeld had invented a crude instrument to measure listeners' responses to music, but he also was interested in judging the impact of journalistic crusades, including those on the radio. Stanton produced a new design for Lazarsfeld's instrument, and by 1940 CBS was using it to study reaction to a wide range of programs, including documentaries.[84]

The Program Analyzer laboratory grouped a panel of listeners around a table. Each listener was given a handheld device with a red button and a green button. A large sign next to the table indicated how the panel should operate the buttons as it listened to a radio program: press red for "Poor / I Don't Want to Listen," press green for "Good / I Want to Listen," and press nothing for "Just Average / I Am Indifferent." The collective responses were recorded on a sort of polygraph from the program's start to its finish. The bigger version of the Analyzer could track one hundred listeners at once and was dubbed "Big Annie."[85] CBS researchers billed it as "a form of jury trial with the listeners as jurors." Based upon their studies, the researchers asserted that educational programs "must be built with real showmanship." Certain content could be dangerous; unlike with a movie or play, "very rarely will anyone continue to listen to a radio program once he has become confused or annoyed."[86]

CBS used Big Annie to test a recording of *The Eagle's Brood* not long after it aired. The researchers reported that "the program was very successful in arousing listener concern about the problem of juvenile delinquency." The panel of listeners were surveyed before and after they heard the program. It was found that 42 percent were "very much more concerned than before" about delinquency, with another 30 percent "somewhat more concerned." Before hearing the program, only 17 percent felt that delinquency was an issue that concerned them personally; afterward, 80 percent felt that way. The overall Program Analyzer score was said to be "exceptionally high for both sponsored and sustaining programs." In the minute-by-minute tracking of listener responses, however, "the outstanding low point" came in the description of the teens awaiting execution in the death chamber.[87]

"I had deliberately written that scene to shock the audience," Shayon would recall. When he asked the researchers whether he should have omitted it, they replied that the decision was up to him.[88] Still, Shayon said that it pointed to "a bone of sharp contention between the psychologists and the artists who said, 'Very often I want the audience to hate somebody.'" There also was the concern that CBS and the other networks were using research more to sell themselves than to improve their programming. Looking back years later at his original partnership with Ernest Dichter, Shayon lamented that more had not been done to "bring together the world of the creative broadcaster with the social scientist" in studying programs' impact on audiences.[89]

Decline

In the short term, at least, the Documentary Unit remained a subject of interest for social scientists. The CBS research department analyzed several other unit productions. It found, for example, that *A Long Life and a Merry One* had been "most strikingly effective in informing listeners that Negroes do not get as good medical care as white people, that there are not enough hospitals and doctors in the United States, and that we are not the world's healthiest country." Regarding *The Sunny Side of the Atom*, 71 percent of surveyed listeners said that they had learned "a great deal of new information" concerning atomic energy.[90] Similarly, a Michigan sociology professor found "almost universal expressions of approval" from Kalamazoo listeners regarding *The Eagle's Brood* and *A Long Life and a Merry One*.[91]

Meanwhile, the Documentary Unit maintained an ambitious production schedule during the second half of 1947. It broadcast two shows on a newly opened set of Abraham Lincoln's papers with the help of a leading group of Lincoln scholars, including Carl Sandburg. In August, CBS forfeited a reported twenty thousand dollars in advertising revenue to air *We Went Back*, which departed from the docudrama format to feature actuality recordings of people living in former war zones such as Pearl Harbor, Corregidor, Normandy, Berlin, and Hiroshima.[92] Cracks in the CBS recording ban widened as 1947 progressed, and Robert Heller was especially keen to use actualities, saying that those in radio had "become too soft, preferring the insulation and manufactured air of our studios to the cold reality of the world outside."[93] *Billboard* called *We Went Back* a "resounding knockout" in showing how the world was "not celebrating the second anniversary of peace, for peace will not have come until its security in the future is certain." In contrast, *Variety* branded it one of the Documentary Unit's "major disappointments to date" and placed a large share of the blame on the use of actualities: "The monotone recording, even of lived-through events, does not add to reality on radio. In fact, any good dramatic actor, benefiting by an intelligent director, could be more effective in driving home the truth."[94]

Whether in response to such criticisms or not, the unit's final two productions of the year reverted to dramatizations and actors. *Fear Begins at Forty* was written by Arnold Perl, the author of *The Empty Noose*. It starred Eddie Albert as a middle-aged man who, after much agonizing, is forced to put his ailing mother in a nursing home and his father in a boarding house. "Well, maybe you were hoping this would be a story with a happy ending," said Albert toward the broadcast's conclusion. Instead, the program had related "hard, unpleasant facts [that exist] because of the way most of us think—or don't think—about old age." The program advocated the expansion of geriatric studies and insurance benefits, the suspension of mandatory retirement ages, the improvement of nursing homes, and the development of senior centers. The second program, *Among Ourselves,* aired at the end of December and told of four real-life incidents during the past year in which citizens had united to protest incidents of racial or religious intolerance. "What they did were some of the successes—among the many failures—of 1947," said the narrator at the end of the program. "Now we shall have another year—another chance—to live AMONG OURSELVES."[95]

As 1948 began, though, major changes were under way at CBS. Edward R. Murrow had stepped down from his executive position the previous summer. There had been rumors of friction between Frank Stanton and himself that Murrow denied, saying that he simply wanted to return to the air as a journalist.[96] Nevertheless, he clearly had been unhappy as an administrator, telling the *New Yorker* a few years later that he "didn't like firing people." Particularly unpleasant had been William Shirer's angry departure from CBS in 1947. He had publicly claimed ill treatment due to his liberalism, thus again bringing vilification from the left upon the network. Beyond such bitter episodes, Murrow believed that he had fallen short of his goal "to revolutionize radio from the inside—make it adult and intelligent."[97] For her part, Ruth Ashton disagreed that Murrow had been a poor executive. "Nobody was greater than he," she said years later. "He cared so much. So he was a great person to have as your boss. But he didn't want to *be* your boss."[98]

Murrow's departure was followed by changes within the Documentary Unit itself. Robert Lewis Shayon would become producer and director of the historical series *CBS Is There*. Without Murrow as a buffer, the tensions between Ruth Ashton and Robert Heller came to a head over an education documentary called *Report Card* that aired in March 1948. Ashton had done some reporting for the program, and as she later told it, Heller "twisted" her work around "so that I didn't even recognize it." She felt that he "had an ideological bias" (presumably a liberal one, although she would not say so directly), "and he tried very hard to inflict it on us, literally inflict it on us." Ashton quit the unit to join CBS's growing television-news operation.[99]

Heller too would leave the unit in 1948 after being promoted to executive producer and second-in-command to CBS's programming chief Hubbell Robinson. Although his new position focused primarily on entertainment shows, Heller was also now extolling the possibilities of documentary on television: "The vistas offered by the infinite combinations of actuality, of films, of cartoons, of animation, of studio dramatizations—these are enough to make a videomentarian drunk with the promise of his new craft."[100] Werner Michel replaced Heller as head of the Documentary Unit. Within a few months, though, it was reported that William Paley now felt that "the one-shot programs and series primarily concerned with getting over a 'message' should be reduced to broader entertainment values designed for greater mass appeal." The Documentary Unit remained active through the end of the decade and produced well-regarded programs such as Arnold Perl's *Mind in the Shadow* on mental health, but the unit's heyday had passed.[101]

Otherwise sympathetic observers argued that the highly touted CBS documentaries had had limited impact. The Michigan sociology professor who had reported "almost universal expressions of approval" concerning the programs noted that the audiences for them were still comparatively small. He also said that stations appeared to air the programs only to "lessen the public pressure for this sort of thing," lacking "the motivations of a true educator or public servant."[102] The critic Saul Carson noted another concern: the "CBS Documentary Unit was never given a regular time and a definite cycle on the basis of which to build an audience." The lack of a regular schedule had been deliberate; according to Murrow, the strategy was "to present this type of program to what might be called an unsuspecting audience," much as had been done with *The Empty Noose*. However, as Carson saw it, the downside of having the documentaries appear at irregular times and intervals was that they typically ended up being "just another catch-as-catch-can entry in the broadcaster's 'public service' dossier."[103]

* * *

One must recognize the truth in Carson's words in assessing the legacy of the CBS Documentary Unit. As was observed at the time, ventures such as the unit were "a partial answer to the insistence of the Federal Communications Commission, enunciated in the so-called 'Blue Book,' that radio devote more and better time to programs in the public interest."[104] According to Shayon, such programs "made a good impression on opinion leaders. Radio could be touted as a public service medium—the losses in commercial time sales were sustainable."[105]

By the summer of 1948, attitudes had shifted at CBS, as Norman Corwin discovered while dining with William Paley aboard a cross-country train. Paley warned Corwin of a growing reactionaryism in broadcasting. "CBS was known as a liberal network, I as a liberal writer, and now he had said 'no place is safe for liberals,'" Corwin recalled of the conversation. Paley by that time also had rejected any notion that CBS should concede a larger audience to NBC in exchange for prestige; the network already had lured Jack Benny from its rival, with other stars to follow. In that same vein, Paley asked Corwin to write for a bigger audience, saying that to do otherwise would not be "making the best use of our talent, our time, and our equipment." In such a climate, programs such as the ones Corwin and Shayon had written would eventually "be thrown overboard," as Shayon described it; they had served their purpose.[106] The radio-reform movement that had given rise to the Blue Book fizzled. The Blue Book itself was viewed as a "dead duck," whereas a similarly reform-minded report of 1947, Llewellyn White's *The American Radio*, made little impression on broadcasters. "It's old stuff and the industry is getting a little tired of it," an unnamed network executive said of White's book.[107]

The networks' investment in research reflected institutional priorities. Robert Heller had seen such research as helping radio make "its richest contribution to the democratic life," but others saw it differently. According to one scholar, the Lazarsfeld-Stanton Program Analyzer "symbolized the marriage of convenience between mass media administrators and the academic community."[108] For academics such as Paul Lazarsfeld, it furthered a social science that later would be sharply criticized for serving commercial ends and for downplaying—if not grossly underestimating—the media's impact on society.[109] For CBS's Frank Stanton, it was "a razzle-dazzle thing to give to Hollywood and the advertisers," as he said years later.[110] The Analyzer's discouragement of that which "confused or annoyed" listeners, including abrupt shifts in tone or place, worked against radio's most creative impulses. Similarly, Ernest Dichter had focused on individual listeners' psyches as opposed to wide-ranging social remedies, and he soon turned his attention to marketing consumer goods.

In addition, criticism could be aimed at the documentaries themselves. Shayon acknowledged that *The Eagle's Brood* was deliberately melodramatic and "flamboyant."[111] It did not match the artistry of earlier docudramas, such as Norman Corwin's *On a Note of Triumph*. Regarding the documentary's delinquency focus, one scholar has since argued that postwar fears about the issue reflected less an actual increase in juvenile crime than they did a concern that the war had undermined the family and public morality; they also were part of a long-standing cycle whereby officials and pundits have periodically deplored mass culture's impact on American youth.[112] Saul Alinsky's community-organizing response to social ills also would come under scrutiny. Just as *The Eagle's Brood* was criticized for not addressing the structural inequities in society, Alinsky was lambasted for not promoting a broader, more truly radical program.[113]

As for *The Sunny Side of the Atom*, it has since been cited as a prime example of the media's cooptation by a government propaganda campaign "to implant in the public mind an image of atomic energy associated with health, happiness, and prosperity rather than destruction," thus helping mute concerns about atomic weaponry.[114] Furthermore, Ruth Ashton, who as a woman was not allowed on the air at CBS, suffered a degree of patronization when Robert Heller proposed turning her story into an Alice in Wonderland–style fantasy. Although she was able to block that with Murrow's help, the resulting program still was criticized for featuring a female narrator whose very presence was thought to undermine the gravity of the subject. That reflected the dictates of what a radio historian has called "a deeply sexist industry in which it was gospel that people did not like and would not trust the female voice on the air."[115]

Given all that, one might overlook the considerable accomplishments of the Documentary Unit. It was in fact a positive industry response to the concerns the Blue Book highlighted, suggesting that such reports had a constructive impact. On the first anniversary of the Blue Book's issuance, the FCC commissioner Paul Walker said that it "definitely had a stimulating effect on public service programs and public service responsibilities of broadcasters throughout the country." It also encouraged network affiliates to air sustaining programs such as the unit's productions.[116]

Likewise, the network's research studies had a positive effect in indicating that the documentaries were succeeding in informing listeners about important public issues. They also pointed to some ways in which the programs could be more effective, much as Robert Heller had hoped. For example, CBS researchers found that "a program which at its very outset strikes a responsive chord, of a personal nature, in the listener has a better chance of holding its audience," a principle that then was applied to future documentaries.[117] Beyond that, it should be noted that Paul Lazarsfeld was not simply an apologist for the broadcasting industry. A 1946 study that he coauthored, *The People Look at Radio,* was seen in some quarters as "a direct repudiation of the FCC's chastisement" of broadcasters in the Blue Book in that it characterized public attitudes toward American radio as largely positive.[118] At the same time, Lazarsfeld observed that "psychologically speaking, supply creates demand," meaning that mediocre programming may have cultivated a taste for more of the same. "Perhaps," he continued, "the taste of the listeners could be 'elevated' and larger audiences obtained if there was a larger supply of more serious broadcasts with a great deal of promotion put behind them." That was precisely what the Documentary Unit had tried to do.[119]

One scholar argues that Lazarsfeld was in fact skeptical about radio's capacity to serve the public interest. Instead, he was among a group of academics who "were concerned primarily not with participatory democracy, but with finding a way for experts to gain effective voices" through the media so that they could communicate effectively with the rest of the populace.[120] In contrast, the Documentary Unit was very much concerned with participatory democracy. Thus did Shayon promote neighborhood councils in *The Eagle's Brood* and call on Americans to clean out "the stagnant pools of ignorance, indifference, [and] injustice,"[121] much as he had concluded *Operation Crossroads* by having Archibald MacLeish say that the collective will should be decided through "talking and listening and thinking" rather than through consuming media. Similarly, *The Sunny Side of the Atom* ended by invoking Albert Einstein's admonition that "only talk between men [and women]" would produce necessary change, just as it also invoked the need for citizens to educate themselves about atomic energy, lest the most important decisions about it be made behind closed doors.

To an extent, such sentiments reflected an overtly progressive agenda. Shayon and Heller, like Norman Corwin, had participated in meetings sponsored by the Independent Citizens Committee of the Arts, Sciences, and Professions and its successor, the Progressive Citizens of America (PCA). "We know that it is not reaction which is on the march and we must stop it," Shayon told a PCA gathering in May 1947. "We know rather that in the truer sense, expanding democracy is on the march and reaction is trying to stop it. And we know where we stand. We want to stop reaction. We want democracy to expand."[122] The Documentary Unit's calls for improved health care and geriatric services as well as for racial and religious tolerance were consistent with such a philosophy, which Ashton felt was eventually pushed too aggressively upon the unit's productions. At times the programs verged on explicit editorializing—which is ironic, given that official FCC policy at the time discouraged such advocacy and that Murrow had unsuccessfully tried to set up the radio equivalent of a newspaper opinion-editorial page at CBS before he resigned his executive post.[123]

In due course, all that would lead to charges of subversion and blacklisting, but the unit's productions were not radical in their politics. Some were hardly controversial at all, as with *Experiment in Living*'s declaration that democracy was superior to autocracy and anarchy. (One of the unit's final productions under Robert Heller's supervision commemorated the start of baseball season.[124]) Institutional priorities again played a role, with *The Eagle's Brood* being discouraged from turning a critical eye on the media and *The Sunny Side of the Atom* promoting the government's interest in a positive image of atomic science. There also was the innate caution of Edward R. Murrow, whatever preacherlike passion he might have displayed on the air in reporting the London Blitz. He was recognized as a liberal, but as an acquaintance said, "Ed never pushes his liberalism beyond a carefully calculated safety point."[125]

Still, a key tenet of Murrow's liberalism—and one that would reappear in his later work for CBS—was that informed discussion is infinitely preferable to ignorant, unreasoning fear. That had manifested itself in Documentary Unit productions such as *The Sunny Side of the Atom*. Those programs earned the approbation of none less than Charles Siepmann, the author of *Radio's Second Chance* and a contributor to the Blue Book. In a year-end 1947 article for *The Nation,* he credited the documentaries for providing citizens with "the facts about the unfinished business of democracy" and showing "us our responsibility for the inhumanity in our society." The article was headlined "Radio Starts to Grow Up," suggesting that Murrow had in one respect succeeded in his quest to make the medium more adult.[126]

The CBS Documentary Unit's experiments with dramatizations and actuality would serve as models for similar experiments at the other networks. It is to that documentary work that we now turn.

4

Home Is What You Make It

If CBS started its Documentary Unit in part to differentiate itself from NBC, the rival network was slow to respond in kind. Writing in the *New York Times* in August 1947, Jack Gould described CBS's documentary efforts as "magnificent topical radio," whereas NBC offered "outmoded and old-fashioned public-service programs" that relegated the network "to a back seat scarcely compatible with its acknowledged stature in radio." However, according to Gould, that seemed poised to change with the promotion of NBC's Ken Dyke to programming chief. Dyke had taken leave from the network to head the military's Civilian Information and Education Section in occupied Japan immediately following the war. His job was to reduce militarism and authoritarianism in Japanese society and to promote democratic values through the country's media. Fresh from that experience, Dyke returned to America with one of his missions being to reinvigorate NBC's public-service programming.[1]

It was not until after the start of the new year that the network's first showcase documentary series debuted. It too was aimed at promoting democratic values with the help of a writer who had honed his craft on a show targeted at stay-at-home mothers.

Lou Hazam and *Living 1948*

Lou Hazam was thirty-four when he joined NBC in 1945, following jobs writing commercials and working for the U.S. Department of the Interior. He was assigned to write the women's series *Home Is What You Make It*, which, he later recalled, was not succeeding when he and three research assistants assumed responsibility for it: "We were permitted to change anything but the horrible title—and did. Indeed, the whole character of the show changed to dramatic documentary; sometimes pure drama."[2] The overhaul of the program actually fit the existing title well, however much Hazam disliked it. In keeping with the spirit of the era's radio documentary, it implied self-improvement and remaking the world for the better while pointing toward historical and contemporary examples.

The first episode under Hazam's oversight, "Heritage of Home," aired in October 1945 and set the tone for what was to follow. "As mankind faces a new atom-born epoch, we invite you to scan the ages with us—and draw upon music, poetry, literature—upon human records and experience—to see how 'Home' has shaped our lives," it began. It pointed out that "home" was the thing most thought about and longed for by those who had fought in the war. "Home is the common denominator from which a greater understanding among nations can grow," the episode concluded. The series would dedicate itself to helping listeners "make your home a better home, and to lay a brick—small as it is—in the foundation of the lasting peace and happiness of the world."[3]

In the following weeks, *Home Is What You Make It* alternated between shows oriented toward what were deemed traditional "homemaker" topics and programs directed more expansively toward the One World–type philosophy that "Heritage of Home" had espoused. Episodes discussed how to care for children of various ages, including one called "The Forgotten Age" on preteens: "Yes, parents can ease their feelings with the thought that the girl Sprig calls 'Dogface' today may be the very girl that'll set him to pleading for the family car, so he can take her out on a date tomorrow!" Other shows stressed the need for citizen action. A Thanksgiving program presented a reverential thanks for the coming of peace, with a onetime defense worker (played by an actor) saying that the same can-do spirit that had won the war could now "put bread in a hungry man's mouth—shoes on a kid that needs a pair of shoes—electric lights in a house that never saw 'lectricity before—[and] give us better health, more education, good jobs, and make life worth living." Another episode, "Your World Neighbors," emphasized that Americans were now "*world* citizens" who should "[p]raise acts of understanding and courage in interracial, international, and interreligious relations wherever you might see or learn of them."[4]

Soon after Ken Dyke assumed command over NBC's programming, Hazam was assigned to the new series *Living 1948*. Hazam later described it as "essentially the same type of program" as *Home Is What You Make It,* but with a key difference: it "was free of any duty to Home Economics subject matter and could scan the whole horizon of man's [and presumably woman's] interest and problems."[5] The new series did operate under one constraint—unlike CBS, which had gradually relaxed its recording ban starting with *One World Flight* the previous year, NBC still rigidly maintained its own ban. Thus Hazam, who later as a television documentarian would shun actors and dramatizations because he thought they injected "falsity," was forced to use them in *Living 1948* in lieu of recorded actualities.[6]

The docudrama format did afford room for creativity, which Hazam had taken advantage of in *Home Is What You Make It,* but the world climate had changed significantly in the three years since he had started work on that show. In 1948, the cold war would escalate through events including the communist takeover of Czechoslovakia, the Soviet blockade of West Berlin and the subsequent Berlin Airlift, and the resumption of the American military draft. The One World sentiments of *Home Is What You Make It* were tempered accordingly in *Living 1948*. NBC said that the new series' mission was "to set Americans thinking more about currently important issues and arouse them to intelligent action." To that end, it drew upon the regular input of George Gallup and his pollsters in discussing how citizens seemed to feel about those issues.[7] Hazam was the primary writer of the series under the auspices of NBC's public-affairs staff, supervised by Wade Arnold. It aired as a sustaining, nonsponsored program on Sunday afternoons. That was not the equivalent of the prime-time slots that CBS had given its documentaries, but it was a regular, predictable slot—at least when the network did not preempt the series to offer time to some of that year's presidential candidates.[8]

The premiere was Leap Day, February 29. *Variety* described it as "flitting from the high cost of living, the New Look in clothes, inadequate housing, unescorted women in bars, and the Marshall Plan to a new Saroyan play and the song called 'Civilization.'" George Gallup was also on hand to say that his polls showed that a majority of Americans were happy, in contrast with France, "where two thirds of the people who were unhappy were Communists." *Variety* commended the program for its ambition but criticized the Gallup interlude as being "a disconcerting lulling-people-into-contentment note" that detracted from the overall impact: "Granted that we are living in the best of all possible lands, the fact remains the issues facing our land require fuller understanding, clear and unbiased."[9]

Living 1948 continued to "flit" from subject to subject in subsequent episodes, with a mix of harder-hitting programs that appeared to heed the admonition to promote understanding of national problems alongside more playful shows suggesting that whatever those problems were, Americans indeed lived in a fine country. The two constants would be the narrator Ben Grauer and either Gallup himself or one of his associates. Grauer was well known to radio audiences as the announcer for the NBC Symphony Orchestra as well as other network broadcasts. He delivered a standard welcome for each installment of *Living 1948*: "*Greetings,* America!" Gallup provided a scientific gloss to the series' probing of the American mood. Although seemingly objective, Gallup's research was not necessarily neutral, as the *Variety* review hinted at the time. One scholar has since observed that his work "was frankly commercial, commissioned by companies that wanted to increase the profitability of their newspapers, magazines, radio programs, or films."[10] NBC was unlikely to see higher profits through Gallup's role in a sustaining series, and it is unclear whether the network paid him for his participation. Regardless, Gallup benefitted from the publicity, while NBC benefitted from its association with the Gallup brand, which as of February 1948 carried considerable prestige.

The series followed its debut with one of the tougher-style episodes, "The Mental State of the Nation." It began by declaring that it was broadcast in "the public interest—*your* interest" and added that it was one of "a new series of radio drama documents reflecting the life of today, to help you reach the decisions that will spell a better tomorrow." (The series would continue to bill itself as a "drama document," suggestive of the grey area it inhabited between fact and fiction.) In dramatizing the nation's mental-health crisis, the program asserted that many public institutions for the mentally ill were "*condemned, over-crowded, outmoded.*" It called for more research, clinics, medical attendants, psychiatrists, and wards, adding that such initiatives depended "*on our ability to work with other men and women to improve the quality of our life as we live it with ourselves and with others!*"[11]

The next week saw a marked shift in tone. "Of Rats and Men" tackled another public-health concern—a massive rat infestation—but impishly told its story through the purported voice of the rat itself. "Do you honestly think that you two-footers running around with your dime-store rat traps and lousy cheese are any match for us?" the rat sneered. Eventually the rat was gassed as part of a public extermination campaign, but not before defiantly vowing that "*rats* shall yet *inherit the earth!*"[12] In response to those initial episodes, Jack Gould praised *Living 1948* for being "intelligent and constructive" but added that Hazam's scripts "leaned a shade to the melodramatic" and that the series' vigorous proclamation about being in "the public interest—*your* interest" was unnecessary. (The series soon dropped those words from its weekly self-introduction.) Gould also questioned the choice of subjects, saying that "the problem of rats" did not seem "the boldest way of approaching America's responsibility in nurturing the democracies of Europe" or other concerns of greater importance.[13]

In fact, *Living 1948* would address international affairs in later programs, again through a cold-war prism. "From a Gentleman in Mufti [i.e., civilian clothes]" dramatized General Dwight Eisenhower's final report as army chief of staff before his return to civilian life. Art Carney, later of *Honeymooners* television fame, skillfully impersonated Eisenhower. Although cautioning against excessive public fear, the general warned that other nations in the future would "be armed with weapons of terrifying destructive power." Hence the nation needed to maintain military readiness, including a substantial infantry; it was all up to "the men and women of the United States determined to retain their liberties and to stand firm against aggression." The Gallup representative William Lydgate reported that a majority of those polled supported an enlarged armed forces backed with higher taxes along with universal military training and a resumption of the draft. A few months later, after the draft had officially been reinstated, Gallup himself appeared on the episode "New Draft, New Army" to announce continued public support for the move. He said that it showed "that in the midst of Living 1948, the American people are firmly committed to a policy of preparedness."[14]

That is not to say that the series adopted an overtly belligerent stance. As the title implied, "As Europe Sees Us" examined European attitudes toward the United States. (The program seemed to draw inspiration from the CBS series *As Others See Us,* an Edward R. Murrow initiative that reviewed overseas press coverage of America.[15]) Piggybacking on reporting by the *New York Times* and *New York Herald Tribune* as well as NBC's own staff, the program concluded that not all Europeans viewed Americans "as white knights in armor charging the black dragon." The cold-war subtext remained; Italians were about to go to the polls, and there were concerns the Communist party might win. The United States was engaged in a "battle of ideologies," the program warned. Nonetheless, Americans should "accept the challenge as world citizens—not as a giant flexing muscles, but with deep humility facing the future with courage and intelligence." In that way, they could "help spell the fate not only of Living 1948, but indeed of decades to come."[16]

Other episodes underscored the "world citizens" theme. "Then and Now" marked the third anniversary of the war's end in Europe by comparing the world of 1945 with that of 1948. The program noted that "peace has a bittersweet taste for many of us" in that "the things for which we fought the war and paid more than lip service are still not secure." It added that Americans should "gird ourselves in statesmanship" even as they again should recognize "that since some nations only understand might," renewed defensive strength was necessary. Similarly, an Independence Day program, "American Self-Portrait, 1948," featured George Gallup saying that although Americans were suspicious of Russia, they did not want war and thought that meeting with its leaders might be a good idea. He also reported that a majority of those polled supported strengthening the United Nations to control the military of all nations, including the United States. "In short, Dr. Gallup," said Ben Grauer, "the typical American, while dissatisfied with the progress of the United Nations, wants to keep on trying and is way ahead of most of its statesmen in the steps he's willing to take to strengthen it?" "So the people say," Gallup replied.[17]

On the domestic front, *Living 1948* presented episodes comparable to the mental-health program that had been among its first offerings. "Plight of Our Hospitals: A Drama Diagnosis" began with a deft satire of a typical "doctor's show" or movie, in which skilled physicians cured life-threatening ailments with heroic efficiency. "Yes, this is the picture many of us have of hospitals: fine services, latest equipment, built and supported by some kind of magic, and readily available to all who need hospital care," said Grauer in his narration. "Now—listen to the *facts!*" Over increasingly agitated music, the narrator listed a multitude of problems: a shortage of hospitals in rural areas, overcrowded hospitals in urban areas, and obsolete hospitals everywhere. The program urged listeners to take advantage of new legislation offering federal matching funds for the construction of new facilities and to support programs underwriting young people in medical school. "The plight of our hospitals, America, is everybody's business," concluded Grauer. "As we said so often during the war, only a strong America—and that means a healthy America—can continue to be a democratic America."[18]

"Danger—School Zone" dramatized a new book by the *New York Times* education editor Benjamin Fine, who had toured the nation's public schools. The journalist was played by the actor John Larkin, best known as the radio voice of Perry Mason in the lawyer series of the same name. In similarly prosecutorial fashion, the journalist in the program declared that "America's public school system is in the midst of the most serious crisis in its history. Our children are being *cheated!*" The prime culprits were a national teacher shortage, along with existing teachers who were woefully underqualified. As for who ultimately was to blame, the journalist asserted, "The only answer

I know is—*you!* You, sitting in your living room listening to this radio program, you are to blame! You and I and our neighbors!" More money was needed for teacher salaries and school facilities. Concerns about cost and government meddling were briefly raised and summarily dismissed. The actual Benjamin Fine (rather more mild-mannered than his docudrama counterpart) appeared at the end: "To make democracy work, it is necessary that we maintain a strong system of free public schools. Nothing less will do. Inferior schooling today means inferior citizens tomorrow."[19]

Not all was amiss in America, of course. A Memorial Day show, "Freedom Is a Home-Made Thing," celebrated self-reliance and individual choice while urging listeners to instill the same values in their children. "The Sun and You" described the many blessings that the sun bestows, including good health. Although the program allowed that overexposure could be a problem, a doctor was heard urging a new mother to "put your baby in the sun as much as possible. . . . Don't be afraid of doing it, even in winter!" "USA—Growing Pains" depicted "Uncle Sam" (played by Will Geer) undergoing a less-than-routine physical examination. While under sedation, "Sam" encountered interlocutors representing various population trends, including a Mae West–style character symbolizing the nation's fertility rate. ("Call me Bertha . . . *Birth* for short," she purred.) Sam boasted that the country had just experienced "the biggest crop of babies you ever saw" in what turned out to be the cusp of the postwar baby boom. "Bertha" retorted with what then was the prevailing view: "War babies. High prosperity. Wartime psychology. More marriages. A temporary situation. . . . The truth of the matter, my bearded friend, is your birth rate's been goin' down and down for *decades*." That trend was fully expected to continue, but America did not need to worry as long as it prepared accordingly.[20]

Regarding women, who had been the target audience of the series' progenitor, *Living 1948* offered a mix of condescension and approbation. "Silver Cords and Apron Strings" marked the centennial of the Seneca Falls women's rights convention. The program lightheartedly thrust Ben Grauer into the role of an insecure male supposedly aghast at the rise of female power and privilege. According to the show, women represented "better biological specimens" than men in that they lived longer, had more physical stamina, could read and write faster, had a larger vocabulary, and were less prone to suicide. Women also controlled much of the nation's money and constituted a growing share of the labor force. "Add that [women] can own property, join unions, own businesses, sign checks, run for office, sport pants in public places—in short, it looks as though they've *won* the battle they started in 1848!" concluded

Grauer, while noting that it did not necessarily mean "that we're a matriarchy, a henpecked nation, or even that women are any happier."[21]

"Home Broken Home" depicted disintegrating marriages' toll through the voice of a divorced woman: "Socially, I'm just a fifth wheel. An extra woman. A bother to any hostess. . . . If I was a widow now, there'd be something hallowed about me. Protected in [men's] eyes. But a divorced woman [is] open prey!" Yet the program attributed the rising divorce rate in part to outdated attitudes: "It is a symptom that marriage must take full account of the social and the economic emancipation of women, the importance of freedom, and individual happiness." Better premarital training and sex education in schools were needed. George Gallup again was present to add that the world situation contributed to the strains on marriage: "The tensions which are a product of Living 1948 call for a very mature, stable human being to make marriage or any other human relationship work."[22]

Gallup and polling themselves were the subjects of "Wisdom in the Streets." The program dramatized the legwork that went into polling while also explaining sampling and other techniques. It labeled Gallup "the Babe Ruth of the polling profession" and enthused about his accuracy in predicting past elections. "Polls represent a new and useful tool of democracy," said Gallup. "By revealing the temper of a people to their leaders, they serve to speed up the process of democracy." Grauer agreed, saying that it was "a thought-impelling commentary on our times that the public-opinion poll can operate only among free people." He reminded listeners that the series had regularly drawn upon Gallup poll results: "For *Living 1948* believes in the eventual insight of the people—believes that the final anchor and security of our way of life is indeed the wisdom of ourselves—the wisdom in the streets."[23]

The streets' wisdom that November proved to be Gallup's Waterloo. He and other pollsters had predicted a decisive victory for the Republican Thomas Dewey in the presidential race, only to see Harry S. Truman win. The *New York Times* observed that "polls had won great influence in national affairs. That influence has suddenly collapsed."[24] The same day that those remarks were published, *Living 1948* poked fun at the man it had so recently lionized. "What Happened?" began with the sound of a telegraph key and the mocking words of messages purportedly sent to Gallup in the election's wake: "Desperately in need of money. Who will *not* be Governor of New York in [19]50?" Gallup gamely played along, saying, "We've taken bows when we're right, so we're ready to take cabbages and ripe tomatoes now." He offered some possible explanations for having been so wrong: some of those who had said they would vote for Dewey stayed home, some of those who had said they were undecided in fact voted for Truman, and some of those who had said they would support the Progressive candidate Henry Wallace voted for Truman. Regardless, Gallup vowed that pollsters would learn from their mistakes: "I hope all thinking people, once they've understood the various problems and factors we must cope with, will not lose confidence."[25]

By the end of its namesake year, *Living 1948* had accomplished what NBC had hoped it would. Before the series premiered, according to *Newsweek*, the "biggest and richest of the networks seemed to be ignoring the postwar documentary." Now, "the documentary had finally come into its own on all four networks."[26] The series would continue in the new year as *Living 1949* (and in succeeding years as *Living 1950* and *Living 1951*). George Gallup and his pollsters, however, would no longer be regular features.

ABC and Robert Saudek

Along with NBC and CBS, the Mutual and ABC networks also were well established in the documentary field by 1948. Under the supervision of the educational director Elsie Dick, Mutual had produced well-regarded programs on subjects including veterans, wartime babies, and adoption. The latter two shows, part of a series called *Your Children Today*, were written by Arnold Perl, the liberal author of a number of documentaries for CBS. Reviewers praised the programs for having "achieved a genuine stature" and for featuring "a topnotch cast and expert handling."[27] Some of Mutual's efforts were less successful, though. *Variety* pronounced 1948's *The Atom and You* to be "the most macabre show ever heard on the air." In an attempt to establish "a new documentary format which would appeal to mass audiences," the program took the form of a game show with a live audience. Contestants were quizzed about aspects of atomic energy, and those who answered incorrectly were required to perform atomic-themed stunts, with one woman hitting a chunk of uranium with a hammer: "In the audible foreground, there was a sound effect of an atomic explosion, an echo of Hiroshima; in the background, there was the convulsed laughter of the studio audience getting a kick out of it all."[28]

ABC's documentaries were more ambitious, with the network hailed as CBS's chief rival in that area.[29] The programs were overseen by Robert Saudek, who directed the network's public-service programming. Variously described as "slight, mild, and totally un-hucksterish" and as "ABC's part-time court jester," Saudek treaded a fine line between being an apologist and a provocateur concerning the radio industry.[30] He defined serving the public interest as literally catering to the interests of as many listeners as possible, saying that if broadcasters "were satisfied to take the easy road of appeasing the nation's handful of intellectuals, we would not, in my opinion, be meeting our obligation as licensees. To disregard the frame of reference of fifty million adults is to be undemocratic." At the same time, he said, "People—millions of people—must know a great deal more than they do about the needs and resources of America and the world," adding that broadcasters needed to exhibit more "courage, both in the selection of topics and in the production techniques which we use." Thus in 1946, ABC aired the reading of John Hersey's *Hiroshima* as well as the similarly themed *Unhappy Birthday* to mark the first anniversary of the atomic bombing.[31]

ABC expanded its documentary efforts the following year while turning its attention to domestic issues. Created during the war after antitrust action forced NBC to divest itself of its Blue Network, ABC was far behind NBC and CBS in terms of resources, stations, and hit programs. Saudek observed that he had "become an expert on working within a limited budget." Still, he also said that such limitations encouraged ABC to experiment: "A limited budget frequently makes the broadcaster think a lot harder about how he is going to present the subject matter in an interesting fashion. To have unlimited funds can make a broadcaster lazy."[32] In its first major documentary of the year, *School Teacher—1947,* ABC employed the docudrama techniques familiar from the other networks. However, it scheduled the program in an unusual three-segment format, with the first two segments airing one night and the last airing the next night. The program was lauded for having "laid bare the weaknesses within the realm of public education," including low salaries and a high burnout rate for teachers.[33]

For its next documentary, *Slums,* ABC took advantage of the fact that it (like Mutual) had no recording ban. Saudek said that the problem of how to address the housing needs of "the underprivileged third of the nation" was "based on rather complex economic facts which somehow did not lend themselves to interpretation by dramatic technique or any customary technique which would arouse interest."[34] The network settled on a two-part approach: a first segment centered around recorded actualities from real-life slum dwellers, and a follow-up consisted of an expert panel discussing potential solutions to the problems that the opening segment had highlighted. George Hicks, who had won notice for his on-the-spot audio recording of the Normandy invasion in France in 1944, was dispatched into slum areas with a wire recorder. The program aired May 20, 1947. As would be true of ABC's other documentaries, it was broadcast without commercial sponsorship.

Slums began with a recording of a man and his two sons playing music on the third floor of a tenement on New York's Lower East Side. Hicks then introduced the program, saying that ABC had taken a recorder into such neighborhoods "to learn—to let people report on themselves—to capture part of a universe on a spool of a thin, shining wire."[35] Apart from New York, Hicks had visited the "notorious shacks" of Boston's North End, Chicago's South Side, and Pittsburgh's Hill District, among other areas. He allowed that the wire recorder could not capture the "broken flights of stairs, the garbage in the halls, the dark corridors, and the strange tenement smell of decay. It doesn't record the inward fears of a young mother, as she hears the rats working in the walls." But it could "hear a baby's cry—the cry of this child who is beginning his life on this same slum block," as a weeping infant was heard in the background.

Interviews with residents followed. A Pittsburgh mother of ten described her flat with no indoor toilet or hot water. She also told of bringing home a newborn who was bitten by a rat and died five days later. A woman in Detroit spoke of slipping through a hole in her floor and being unable to get her landlord to fix it. Another woman talked about being afraid to let her children play in the streets and about how she saw no point in decorating her apartment: "I don't see where curtains would make it look any better." Hicks said that "over a million people" lived in such quarters, with "holes in the floor and ceilings that let the summer rain drip through. Yet human beings live in these; they are there tonight. Only their voices have been carried away on wire."

Slums *did* contribute to society, said Hicks: they produced a quarter of Chicago's delinquent youth, half of the delinquents in Richmond, Virginia, and more than half of the delinquents of Philadelphia. They also contributed a disproportionate share of the cities' adult criminals and fire calls. Hicks took his equipment to a neighborhood bar in New York, where he recorded young people's hopes of fleeing such conditions for a better life. He said that one young woman seemingly "knew that were she to fail, her youth would be gone, the freshness would leave her, the gabardine would grow old and worn, and she would sink back for good into the life she so wanted to escape."

Another person told Hicks of dreaming of living on Long Island in a house with "seven or eight rooms [and] a nice big back yard for the kids to fight and play in." Hicks in his closing narration warned that the worst possible outcome would be if these young people's children were themselves condemned to live in slums: "And so, the reason for our [program's] dedication—to an unborn audience—to the children of a future generation, with the hope that the words you have just heard may never come from their lips—that the boundaries of their lives may extend beyond the pushcarts and train tracks—that the fly-breeding air shaft and the rat-ridden dumbwaiter may never be a part of the furniture of their homes, and that the dark corridors of Slums USA may not be even a memory!"

The wire-recording segment of *Slums* was then followed by the panel discussion with three national authorities discussing the comparative advantages and disadvantages of public versus private housing. For Jack Gould of the *New York Times*, the two-part program "represented one of the most provocative and intelligent presentations of a public issue that radio has made in recent years." He was especially taken with the recorded actualities, saying that they demonstrated "radio's own obvious capabilities as a journalistic medium" while also showing that "the only effective spokesman for the 'common man' can be the man himself." Allowing residents to speak for themselves "made a much more effective indictment of the slums" than any dramatized approach could have offered, with "persons never before heard on the radio [having] had their say from coast to coast in a democratic broadcast." In sum, said Gould, "If broadcasters in the future could emulate this instance of true fulfillment of a network's responsibility to the public, 'blue books' criticizing radio would be a thing of the past."[36]

Variety's review was more measured. It lauded ABC for presenting the program in prime time and "for completely sidestepping all subterfuge and directly charging the people with their duty to take care of public health and security to strengthen our democracy." It also raised concerns that would surface again later regarding similarly themed public-affairs documentaries: the actuality segment was marred by "the failure to strike a dramatic note, regardless of the miserable realities unfolded by the slum dwellers," whereas the expert-panel segment "started off with a stuffed-shirt atmosphere of words without meaning." Although the panel discussion eventually grew more lively, it still could have done more to ensure that the specific problems raised by the actuality segment were "eloquently correlated to provide the listeners with the fire necessary to spur them into action."[37] *Variety* might have contrasted *Slums* with CBS's *The Eagle's Brood,* which had aired a couple of months previously and which *Variety* had praised for presenting "a community plan for attacking the problem from the bottom, by the individual, by one neighbor talking to another and doing something."[38]

Regardless, *Slums* fulfilled Robert Saudek's desire for innovation in production techniques, as would also be true of ABC's third major documentary of 1947. In the place of wire recorders, expert panels, and rat bites, *1960?? Jiminy*

Cricket! substituted song, gaiety, and Donald Duck. Lou Hazam wrote the script as a freelancer and stressed many of the same themes that would become familiar in his work for NBC. The program presented a novel concept, to say the least: it transformed an eight-hundred-plus-page report from the Twentieth Century Fund titled *America's Needs and Resources* into a full-fledged Walt Disney musical. Disney loaned characters and songs to publicize its new animated feature, *Fun and Fancy Free*.

The radio show aired that September. It started with Jiminy Cricket (voiced as usual by the actor Cliff Edwards) singing the title tune from the Disney movie with lyrics adapted for the broadcast:

> Now Frisco, Buffalo, New Orleans, why, folks should be in clover
> You're gonna hear what 1960 means before our show is over
> So if you really want to see just how happy you will be
> Climb right on my song and come with me
> 'Cause tomorrow's full of fun and fancy free![39]

As the song implied, the program drew upon the economic report's findings to show Americans the wonders that awaited them in the year 1960. Grumpy of the Seven Dwarves played the skeptic. "I don't see what all this 'fun and fancy free' business is about," he groused. The newspapers were full of strikes, inflation, war, and other woes, and Grumpy could not find an adequate place to live or save any money: "'Hail 1960'—*fiddlesticks!* I say we're on the downgrade. We're hittin' the skids! And the whole dangblame country's goin' plum to *pot!*"

"Grumpy, I think you're just the man we want to see," said Jiminy. There was another side to the picture, showing that war and depression were not inevitable: "Keep your eye on that side of the picture for a little while. Come along with us, Grumpy, and see if this isn't the way you'd like our country to be." With that, Jiminy, Grumpy, and Donald Duck were off to look at America as it could be in 1960, a land of improved labor relations, shorter work weeks, more leisure time, increased productivity, and enhanced use of natural resources. In agriculture, science would eliminate crop disease, as underscored by a lugubrious basso-profundo ditty sung by a worm ("I'm the victim of an adverse circumstance.... A worm like me just doesn't have no chance"). As for slums and the housing shortage, they too were a thing of the past—that is, "if you want it that way, if you put out just a little bit more oomph to deliver the goods," as Jiminy put it to Donald and Grumpy. They boarded a magical rocket that took them to the city of the future. There were superhighways, factories moved to the suburbs, and new housing developments with stores and theaters nearby. The slums had been replaced by green space, fresh air, and happy children singing "The Farmer in the Dell."

"From where we sit, 1960 looks like the promised land," said Jiminy by way of conclusion:

> But there are still some big ifs. If there's no war, and if nobody gets trigger-happy with atomic bombs. If our statesmen practice statesmanship and use their mandate with wisdom. And of course if a major depression doesn't paralyze the world. And there's another footnote: We haven't mentioned any other country because this has been the story of America in 1960. But the welfare of the other countries [is] our welfare too. If our land is to be full of peace and plenty and happiness [thirteen] years from now, it'll be because we've reached out a helping hand across the seas where words like needs and resources have a very different meaning. It's worth working for, don't you think? It may not be perfect, but by jiminy, it's worth *working* for!

Variety declared that *1960?? Jiminy Cricket!* "wasn't half-bad," especially considering the skepticism with which many in the radio industry had viewed the project and the fact that the program had necessarily "skipped over at least eight hundred pages" of the 812-page report that was its nominal source material. "Saudek and ABC ought to take gambles like this more often," the review concluded. "So should some others."[40]

V.D.: The Conspiracy of Silence and *Communism—U.S. Brand*

Saudek upped the ante at the start of 1948 by announcing an ambitious schedule that was seen to "accent once again the keen rivalry between ABC and CBS in the documentary department," according to *Variety*. In the works were reports on birth trends, the Marshall Plan, a "satire on world affairs" loosely following the format of the Jiminy Cricket program, and a three-part series on communism.[41] In April, however, ABC announced that Saudek had "pushed aside plans for a half a dozen other documentaries" to present a different program: an exposé of venereal disease.[42]

V.D.: The Conspiracy of Silence had had its genesis the previous fall when Erik Barnouw was visited by the Columbia University epidemiologist E. Gurney Clark and T. Lefoy Richman of the U.S. Public Health Service.[43] Barnouw was a radio instructor at Columbia, and Clark and Richman invited him to bid for a federal grant to create a series of educational shows about syphilis. Barnouw won the grant, having been the only bidder. He recorded several fifteen-minute test programs in different genres, including soap operas and "hillbilly dramas" with songs by the country-music star Roy Acuff. The idea was to appeal to as many different listeners as possible. Before proceeding further, it was decided, given the sensitive subject matter, to seek the approval of Columbia's new president, General Dwight Eisenhower—the "Gentleman in Mufti" had entered academia, if only temporarily. Barnouw and Clark played a couple of the test recordings for Eisenhower, who responded enthusiastically: "I'm *delighted* we're doing something that isn't *way up* in the academic clouds!"[44]

There was another potential obstacle: would radio stations agree to play the programs? Columbia brought several station executives to New York, where Barnouw briefed them on the project. They told him that it would help greatly if one of the networks agreed to air it. After NBC and CBS both turned down Barnouw, he approached ABC, which he said "was considered the also-ran network." As it happened, that worked to his advantage. Robert Saudek was immediately receptive, seeing such controversial shows as raising his network's profile. Furthermore, the programs on communism that he had planned were running into difficulties and would not be ready by April as originally planned. Thus ABC scheduled an hour-long documentary on venereal disease to air at the end of that month, provided that Barnouw could produce an acceptable script quickly. He wrote it in a week.[45]

Final preparations for the program took place amid growing apprehension. During the war, Catholic groups had pressured the broadcasting industry into killing a radio information campaign about syphilis, calling it immoral. Now ABC was hearing similar protests. "Saudek kept dropping in on rehearsals to reassure himself," recalled Barnouw.[46] The broadcast took place as planned in prime time on April 29, 1948.

George Hicks again narrated. He began by stating the "simple, brutal fact": one out of ten Americans had syphilis or gonorrhea. Some were "innocent" victims: wives infected by husbands and infants infected by mothers. Venereal disease had to be eliminated. The program would tell of medical advances, but it also would "serve to remind you of the spiritual enlightenment, and the renewed respect for the laws of God and Man, which must come to pass, if V.D. in America is to be relegated to history." Similar to what he had done in *Slums,* Hicks used a wire recorder to interview real-life patients with venereal disease. A woman was heard describing her shock upon discovering that she was infected. She was being treated and cured, but others were not, according to a nurse (played by an actor): "[W]e are held back by ignorance—wide public ignorance. This ignorance is dangerous. It could be corrected."[47]

The program then shifted to full-fledged dramatization, telling of a young couple named Kitty and Herb. Their marriage was disintegrating after only two months, with Kitty finally walking out on her bus-driver husband: "I've been doing a lot of things that probably would surprise you! I want some fun out of life." Unfortunately for Herb, Kitty left him with a "souvenir," as George Hicks's narration described over a "brooding background" of music: "A souvenir that, already, has traveled through his bloodstream to all parts of his body. A souvenir he knows nothing about, but that may be with him a long time. Its name is syphilis."

Herb developed a sore and then a rash that both came and went, while Hicks related that undetected, untreated syphilis could lead to blindness and insanity. The advertisements inside Herb's bus hawked palliatives for all manner of ailments and conditions, but nothing remotely connected to syphilis. One could find little on the radio or in the newspapers either: "Because Herb is the victim of a tradition that says: 'Don't mention that word.' Herb—and a million others. . . . They are the victims of a conspiracy that goes back for centuries—a conspiracy of silence."

There followed a historical overview of venereal-disease treatments, which had proven largely ineffective until the 1943 introduction of penicillin. Then the dramatizations resumed with the story of "Jack Burrows, son of the manager of the town's leading supermarket," who was heading off to college. In short order, he meets the peripatetic Kitty on a bus, shares a brief interlude with her, and contracts syphilis himself. Afraid to tell his parents or college officials, he consults a quack doctor, who for a hefty fee gives him a "secret cream." Such phony cures were all too common, said Hicks.

The program came full-circle with a demonstration of so-called contact tracing via Herb, Kitty, and Jack. Herb had remarried and "settled in a beautiful little farm in a fertile valley, miles from anywhere," but he had unknowingly infected his new wife. A contact tracer assigned to track down possible disease carriers steered them both toward successful treatment with penicillin, allowing them to have a healthy baby. As for Kitty, who was about to marry for the third time despite not yet being out of her teens, "a required premarital blood test kept her from carrying the results of her shabby past into her new marriage." However, Jack and many others had yet to seek the proper treatment. "We must help them find themselves," a real-life public-health-service official said. Broader change was needed beyond that, according to the nurse who had been heard at the start:

> In the public health field we know that people like Kitty, who can leave a chain of infection from one city to another, are in most cases the products of broken

homes.... They are failures not of health departments alone, but of society. They are a challenge not only to doctors but also to statesmen, teachers, and churchmen, to bring together their knowledge and understanding, and help us face the problems of our times. Problems of education, ethics, and religion. There are around us *many* barriers of ignorance and silence. But with faith *and knowledge,* we know we can move ahead.

The V.D. project proved a boon to Barnouw and ABC. It helped Barnouw procure grants for additional projects and even briefly gave him a hit song in certain cities. He and the folk musician Tom Glazer wrote "That Ignorant, Ignorant Cowboy," which told of the regrettable consequences of a liaison between a cowpoke and a "girl called Katey." (The song, part of Columbia's ongoing V.D.-education campaign, did cause the university some consternation when the father of a Barnard student complained that it had ruined his daughter Katey's life.[48]) ABC basked in letters of praise from teachers, clergy, and others, as well as reviews calling *V.D.: The Conspiracy of Silence* "a notable and important advance in the educational use of radio" that had met "the real test of honest journalism,"[49] while also proving that ABC was "continuing a policy of courageous usefulness in the public service field."[50] The program had allayed Robert Saudek's apprehensions and done all that he could have hoped it would do.

In the meantime, though, his planned documentaries on communism still faced formidable obstacles. *Variety* reported in May that Saudek was having trouble locating a writer: "He feels that the series must hew closely to an unbiased viewpoint, but finding a scripter who is neither pro-Communist nor anti-Communist is turning into a search." The *New Yorker,* never shy about needling the radio business, said that broadcasters seemed forever determined to unearth "the first-rate writer with the third-rate mind," adding, "That ABC, after a prolonged and most diligent search, failed to uncover a script writer without an opinion of some sort on Communism is, we suspect, the highest tribute that radio script writers will be receiving this year."[51]

In reality, by the time the *New Yorker* story appeared, ABC had found a writer with pronounced views on the subject. Morton Wishengrad was best known for scripting the NBC religious series *The Eternal Light* and the wartime radio docudrama *The Battle of the Warsaw Ghetto.* In an autobiographical note near the end of the war, Wishengrad jokingly said that his pre-radio career had included stints as "a very wild right-handed pitcher, an usher at the New York Paramount Theatre, an errand boy for a watch case company, a confirmed distributor of leaflets for the Young People's Socialist League, [and] a loudly incoherent street-corner speaker."[52] If he was a onetime socialist, he also had long been a vigorous anticommunist. He had worked as educational director for the International Ladies' Garment Workers' Union and supported the union head David Dubinsky's campaign against communist influence in the labor movement.[53]

Wishengrad's acceptance of the ABC assignment drew the wrath of the communist newspaper the *Daily Worker*, which declared that Wishengrad was "just the man to knife labor in the back." As for the upcoming programs that he would write, the newspaper said that the "word 'documentary' is intended to give credence and authority to what will doubtless be an inaccurate, slanderous, and distorted series."[54] Condemnation from such a source may have irked Wishengrad, but criticism from left-liberal colleagues in radio particularly stung. The writer-director Anton Leader scolded him for having "'joined the ranks of active cranks, neo-fascists, sensation-seekers, and red-baiters.'"[55]

Leader's comments came at a difficult moment for Wishengrad. The original plan had been to do three programs on communism. Now there would be only one, but the writing had dragged on well into the summer of 1948, as Wishengrad clashed with Saudek over the script and compensation for it. On July 1, Wishengrad wrote the ABC executive requesting an extra fee for rewrites, which Saudek refused. Four weeks and multiple drafts later, a time during which Wishengrad was unable to take on new script projects for pay, he pressed the issue again: "I have devoted seventeen full days to conference and new writing. . . . It's embarrassing to write [you] a second time; but it looks like I shall be both embarrassed and broke anyway; and so this letter."[56]

In that frame of mind, Wishengrad responded to Leader's criticisms in a separate letter. After describing his fatigue and frustrations regarding the script, he wrote that "had I known it would have turned out to be a career instead of an assignment, I wouldn't have gone near it." Nevertheless, he defended his participation in the program. "I think that Communism is dangerous only to the liberals, of whom you are one," Wishengrad told Leader. "It takes a liberal and makes a shill out of him. It takes sweet, good, humane, unselfish people who have a compulsion to work for social improvement and it confuses them to such an extent that they become unconscious apologists" for the evils of the Soviet Union. He concluded by saying that if he could get liberals "to start actively working for the world they believe in, instead of defending a world they hate, then I will have done something, won't I?"[57]

Communism—U.S. Brand was finally ready to air on the evening of August 2. What was said to be the American Communist party oath was quoted at the start: "I pledge myself to rally the masses to defend the Soviet Union [and be] a vigilant and firm defender of the Leninist line of the party, the only line that ensures the triumph of Soviet power in the United States." Over ominous-sounding music, a narrator (the actor Norman Rose) declared that the program was being "produced in the public interest by ABC so that Americans may better understand and evaluate the meaning of the Communist Party in a free America." The narrator said that the program would deal in documented fact to be underscored by aural "footnotes"—literally, whenever a damning piece of communist dogma was quoted, such as the party oath heard at the beginning, it would be followed by a booming voice saying "*footnote!*" and then giving the exact source and page number from which the quote was taken. "The evidence will be ours," said the narrator. "The conclusions must be yours."[58]

The hourlong program told of the lamentable life and career of "Phil Blake," played by the actor Joseph Julian. (The show was presented as straight docudrama with no recorded actualities.) Unable to find work during the Depression, shunned by his father, and having no other friends, Phil falls in with the Communist party, after initially being recruited by a newsstand operator who pressed a free copy of the *Daily Worker* upon him. He becomes the tool of a menacing party operative known only as Comrade Howard ("I've been a member of the Communist Party since it was organized in 1919—I don't trust my own *mother*"). Phil is assigned first to engineer a takeover of the local chemical workers union, which, according to the narrator, to that point had "lived in the great tradition of American democracy" by following the "overwhelming majority of American unions" in rejecting communism.

Close to his goal, Phil is abruptly pulled off the job. There has been a change in party line, says Comrade Howard: "There's nothing inconsistent about calling Roosevelt a social fascist on Monday and a friend of labor on Tuesday. If necessary, we'll call him a social fascist again on Wednesday." The duty was always to the directives of the Politburo in Moscow, as underlined by the repeated chant of a chorus of communist voices against a martial musical beat: "*We do not question!*" Phil's new task is to organize a series of party fronts by enlisting the support of unwitting liberals. One such person, a "Dr. Whiteside," finally discovers that he is being duped and confronts Phil in words echoing Wishengrad's letter to Anton Leader:

> You take a person who thinks of himself as a liberal and you make a sucker out of him. You take words like "democracy" and "progressive" and "liberty," and you twist them into meaning their opposite.... Some people may think

you're a menace to the capitalist system. I don't think so. You're a menace to *me*. You're a menace to all those who call themselves liberals. You take good causes like race equality and decent housing and civil liberty and you use them for your own ends, just as a come-on for the Soviet Union. . . . I wish I could get up on a soapbox and say to the liberals, go ahead, fight your battles. Go ahead and God bless you. But fight without the help of the communists. Because the thing about a communist is that you can't trust him.

Indeed, Phil eventually disappears under mysterious and implicitly violent circumstances. The lesson was clear, according to the concluding narration:

Our footnote would tell us that only seventy-four thousand Americans hold party cards and that some others hold with party beliefs. The danger lies not in their numbers, but in our own shortcomings. There are things that are wrong in America. We shall approach them with clean hands. We shall correct them our way. If there are inequities, we must work to remedy them. If there are poverty and squalor, prejudice and pride, we must face up to them. But if like the improvident shepherd we allow ourselves to slumber unmindful of our task, then like the shepherd we may awake to find that our flock is gone. For one thing at least is true of communists everywhere—*they never sleep.*

Once more, *Communism—U.S. Brand* received rave reviews that seemed to vindicate Saudek's faith in the project and his choice of Wishengrad as writer. "In a day when the liberal cause is beset by unprecedented doubts and confusion, Mr. Wishengrad took a firm stride forward by showing how that cause has been deliberately sidetracked by Soviet-inspired cunning," wrote Jack Gould in a Sunday piece that included Wishengrad's picture. Gould added that it "was an accomplishment of which both he and the ABC chain have every reason to be proud." *Time* said that the program had "achieved a notable success," whereas *Newsweek* noted that the national response had been so favorable that ABC had scheduled a repeat performance. *Variety* pronounced the show "excellently produced, skillfully enacted, vigorously executed," but cautioned that it "had made a strong, but possibly not watertight, case for the prosecution. The defense isn't apt to be heard from."[59]

Granted, it was not heard from on ABC, which denied the Communist party's request for equal time. However, the *Daily Worker*'s radio critic Bob Lauter (who had predicted that the program would be "inaccurate, slanderous, and distorted") attacked *Communism—U.S. Brand* in three consecutive columns. Among other things, he charged that the purported party oath repeated several times during the show had not been used for years, and its reference to "Soviet power" pointed toward a proletarian-oriented state, not one subordinated to a foreign nation. Likewise, a Lenin quotation attacking democracy was taken out of context; it referred to corrupt bourgeois democracy and not proletarian democracy. The discussion of "fronts" ignored the many instances of communists, liberals, and progressives openly and proudly joining in common cause. In brief, "[T]he program exhibited the high moral standards of a goat."[60]

Again, it was the criticisms of those whom Wishengrad had heretofore seen as sympathetic that especially rankled him. Saul Carson was radio critic for the *New Republic*. Under the editorship of Henry Wallace, the magazine had been outspoken in supporting a reconciliation with the Soviet Union. Wallace had recently left the *New Republic* to run for president under the Progressive party, and he was backed by many on the left, including some communists. Carson himself had supported left-liberal causes such as the Republican side in the Spanish Civil War, and in August 1948 he was about to depart for Poland and the World Congress of Intellectuals, a meeting that attracted many sympathetic toward the Soviets.[61]

Carson had been careful not to condemn the ABC program prior to its airing, saying, "All I know is that it is being written by a radio scripter whose skill I greatly respect."[62] After the broadcast, though, Carson went full bore after the program and its author. "One did not expect a network owner to line up with the Left," he wrote. "One might have expected it to give at least a clue to the other side of the picture." Instead, ABC had served up a straw person to be "pummeled from the first gong, torn apart by every technique known to radio and then pointed to as a miserable weakling." As for the show's writer, said Carson, "I don't see how Wishengrad can ever hold up his head again."[63]

Norman Rosten's remarks were pithier and even more personal. The radio writer, who would declare his antipathy toward "'poverty, war, fascism, corruption in high places or low,'"[64] also held a sardonic attitude toward broadcasting, having declared that "as long as 99 and 44/100 percent of radio is commercial, it will remain an artistic and cultural vacuum."[65] That *Communism—U.S. Brand* had not been commercially sponsored mattered little to him. "Heard your opus last night," he wrote to Wishengrad. "I think you're too good a writer to go in for soap opera. It ain't worth it—not even your picture in the Sunday *Times*."[66]

Wishengrad composed a furious response to the *New Republic* concerning Saul Carson's notice, saying that when Carson had favorably reviewed *The Battle of the Warsaw Ghetto*, "he did not chide me for failing to present the case for the Nazis; nor did he propose that ABC in its recent documentary on venereal disease prove its objectivity by stating the case for syphilis." In the end, Wishengrad decided not to send the letter, telling Robert Saudek, "I don't like a personal exchange and so I've decided to forget it."[67] Concerning Norman Rosten, Wishengrad had no such compunctions. "'Soap opera' implies that I am a literary hack and that the facts and the scenes contained in *Communism—U.S. Brand* are pure fiction. That isn't so," he wrote to Rosten. "The *Daily Worker* called me a hired pen of Wall Street—somehow I resented it less."[68]

An increasingly vitriolic exchange of letters between the two men followed over the next month. "[T]he plight of the seduced liberal leaves me cold—and is he *always* used by the communist?" Rosten wrote to Wishengrad. He said that the program "was a failure as I heard it. Not without an ironic moral: the communist uses the liberal, the corporation (ABC) uses Wishengrad, and everybody is happy including J. Parnell Thomas"—the latter being the chair of the House Committee on Un-American Activities (HUAC) that had investigated communism in Hollywood the previous year. Wishengrad retorted by branding Rosten's allusion to Thomas a "smear" and added, "It comes down to this: I'm a liberal opposed to the CP [Communist party] line. I think you're a liberal who isn't opposed." "No point in arguing," Rosten said in his next letter to Wishengrad, before escalating the argument: "Truth is very often what one believes, and belief is far more complicated than logic. . . . I suppose history will continue its own way, and I suspect the rewards of the communist-hunting liberal will be bitter and brief."[69]

So it continued: "Your most recent letter is an extraordinary theological document. So is the moral posture you assume," Wishengrad wrote to Rosten. "Now you're getting trigger happy," Rosten fired back. "And really, you must get used to the idea that other liberals may not see eye to eye with your theories or documentation. Don't be so dogmatic; you're beginning to sound like a communist."[70] In response, Wishengrad composed a point-by-point rebuttal of charges that Rosten had made in his previous letters, including his original "soap opera" jibe: "Intended by you to be insulting. It was." Then Wishengrad apparently decided that enough was enough. He put aside his original letter and wrote a new one in its place: "Dear Norman, When I undertook the script, I knew I would be assailed. I knew in what manner also. . . . I knew that the script would become a species of political litmus paper. And so now I'm quite anxious to agree with something you wrote in a recent letter, 'We've got different definitions for the same thing, as well as different colored glasses.' Suppose we leave it at that."[71]

"Dear Mort—OK—let's leave it," Rosten replied in a handwritten note. "I hope I haven't sounded persecutory—I didn't mean to be." He concluded with a bit of radio-script shorthand in an implicit nod to the two men's shared profession: "With history in the B.G. [background], and very loud, I wonder just how much of our dialogue is heard anyway."[72]

CBS Is There

Amplifying history had become the central preoccupation of Robert Lewis Shayon by 1948. As auteur of *CBS Is There* (eventually known as *You Are There*), he was producing perhaps the most audacious experiment in postwar audio documentary.

The series had originated the previous year, soon after Shayon created *The Eagle's Brood*. Goodman Ace, who came up with the idea originally, was best known as writer and costar with his wife Jane of the radio series *Easy Aces*; he also oversaw CBS's comedy programming. His new series idea was thus a change of pace for him. It would dramatize historical events by having present-day CBS journalists "cover" them as they would any other major news story. Shayon recalled how Ace conceived the first episode: "Goodie took [his] cigar out of his mouth, and said, 'How about Abraham Lincoln'—he closed his eyes—'at Ford's Theater'—and waved the cigar gently—'CBS Is There.'"[73]

Shayon eagerly embraced the project, and with help from Ace, he wrote and produced a pilot recording of the Lincoln program by the start of April 1947. *Variety* reported at the time that the pilot had "already excited considerable enthusiasm and sponsorship nibbles" and that "CBS is of the opinion it's got the sock educational format of the year."[74] However, CBS executives, including Davidson Taylor and Edward R. Murrow (then still in charge of news and public-affairs programming), did not schedule the new program to air. Murrow told Ace of a serious reservation: "We think actors should be used to play newscasters. Real newscasters may lose their authenticity if they appear in this program." There also was a lengthy debate for reasons unknown over whether the series should be called "CBS *Is* There" or "CBS *Was* There." Without Shayon's knowledge, Ace played the recorded pilot for the *New York Herald Tribune* critic John Crosby, who praised it in his column: "Besides providing a painless history lesson, it's an exciting program."[75] According to one account, that still was not enough to sway CBS executives. It was only after William Paley himself endorsed the new series that it made the network schedule, debuting as a summer replacement series beginning in July.[76]

The new series featured a signature opening. A scene from history would be established via appropriate sound effects and "on scene" narration from a CBS reporter, most often John Daly. Then the narration would give way to a stentorian, echo-slathered voice that announced the date and place before exclaiming, "*CBS Is There!*" (Announcing the date multiple times during each broadcast not only enhanced the drama, it also reinforced that the program was in fact a fictionalized reenactment, reassuring skittish CBS executives who feared a repeat of the alarm surrounding Orson Welles's 1938 *The War of the Worlds* broadcast.) Each episode was said to be "based on authentic historical fact and quotation," with "all things are as they were then, except for one thing"—the echo returned—"*CBS Is There!*"[77]

The debut episode on July 7, 1947, established the template. Daly was heard outside Ford's Theater on the evening of April 14, 1865, over the sound of a happy crowd celebrating the end of the Civil War and cheering President Lincoln's arrival for that evening's performance. The microphone captured a band playing "Dixie" at the president's request. "It's his policy of forgiveness and reconciliation with the South," said Daly. The CBS reporter lugged his microphone through the crowd ("I'm sorry . . . watch the cable!") for a brief interview with the great man himself. "Some think I do wrong to go to the opera and the theater," said President Lincoln. "But it rests me. . . . A hearty laugh relaxes me, and I seem better after it to bear my cross."[78]

Inside the theater, Don Hollenbeck was stationed with a microphone at the back of the house, while Daly assumed position in a broadcasting booth opposite the president's box. In the midst of a running commentary about what was happening on stage, Daly mused aloud about what the president might be thinking: "We've had a hard time since we came to Washington, but the war is over, and with God's blessing we may hope for four years of peace and happiness." Then a shot rang out followed by a commotion in the president's box. "Something is terribly wrong—this is not part of the play!" cried Daly. A wail from Mrs. Lincoln confirmed that the president had been shot. Daly hurriedly switched over to Hollenbeck, who was caught up in the confusion following the shooting ("I guess we're on the air here—I'm trying to get through this crowd to find out what's going on. . . . Watch the cable, please!"). Hollenbeck soon established that John Wilkes Booth was the assassin.

The program then cut between Daly describing Lincoln being carried to the Peterson House across the street and Ned Calmer and Quincy Howe in the CBS newsroom reading bulletins about the stabbing of Secretary of State William Seward that had also just occurred. Daly somberly reported that the president was "sinking swiftly." He interviewed the navy secretary Gideon

Welles, who branded Booth a "lunatic" and "known secessionist." Welles allowed that he did not know the "mind of the South" but added, "I predict that when the men who fought with [General Robert E.] Lee in uniform under a flag and according to a clean code hear of this day, they will shrink with horror from this vile deed. Assassination is not an American tradition."

Daly observed that news of the shooting had spread around the world, as indicated by radio messages from overseas accompanied by the shortwave static and interference familiar to 1940s radio listeners. Britain's Queen Victoria sent words of comfort to Mrs. Lincoln, and Leo Tolstoy paid tribute from Russia: "The greatness of Napoleon, Caesar, or Washington is moonlight by the sun of Lincoln. His example is universal and will last thousands of years." Over the weeping of the throngs outside the Peterson House, Daly delivered his own eulogy for the dying president: "Abraham Lincoln will live on in the union he has saved, in the freedom he has given, in the dreams he has dreamed, in the vision he has seen." Finally came the announcement from Edwin Stanton, delivered personally to the CBS microphone—"Now he belongs to the ages."

CBS Is There was originally slated as only a summer replacement series and ceased production in August 1947, after only seven episodes, but widespread support from listeners and critics prompted the network to reinstate it to its fall schedule. CBS said that it had received thousands of letters and phone calls demanding the program's return, and reviewers called the series "an important educational radio feature" that was suffused with "amazing tension" and that constituted "moving and wonderful radio."[79] Inevitably, the contrivance of placing contemporary radio journalists and technology (including mobile units and portable tape recorders, which were just coming into use) at events from the distant past invited some derision. The *Atlantic* published a parody headlined "Eardrums along the Mohawk," whereas Bernard DeVoto dismissed the Lincoln episode in *Harper's*: "That wasn't terror in Ford's Theater, it was a sound-effect."[80]

Shayon would express disappointment in DeVoto's review, but he never apologized for the elaborate production techniques that he and his crew brought to the series. "We were all technically very ambitious and considered ourselves virtuosos," he later said. "We sneered at a one- or two-mike show—you had to have eight, ten mikes."[81] They took over a large, older CBS studio in New York and placed the microphones and actors everywhere from the basement toilets to the tops of fifty-foot ladders to achieve particular effects. The writers also conducted extensive research, with one dispatched all the way to Haiti to gather background for a program on Toussaint L'ouverture. Shayon cowrote most of the early episodes in addition to producing and directing the series, which aired live each week. He estimated that each show required six to eight weeks of work with up to fifty cast members featured in each.[82]

In addition to the meticulous production and attention to historical detail, the use of real-life CBS newscasters gave the series what one critic called its "startling reality."[83] Along with Ned Calmer and Don Hollenbeck (who after a frantic-sounding debut in the Lincoln show reverted to the more stolid presence for which he was usually known), *CBS Is There* featured Ken Roberts, Harry Marble, Richard C. Hottelet, and Douglas Edwards, among others. Quincy Howe provided analysis, as he had done in the Lincoln program and as he did in real life for CBS; he also was a historian in his own right and rewrote much of his own dialogue with Shayon's blessing. George Fielding Eliot, a CBS military analyst, served in the same capacity for war-themed shows, such as one on Gettysburg set in 1863. Whatever Murrow's concerns regarding the journalists' participation in the series, Shayon later recalled that they "loved their roles.... Because we always built situations according to the then-practiced rules of modern news coverage, the CBS newsmen thought of themselves as 'newsmen'" and not actors. Only Murrow himself remained a holdout, politely declining without explanation Shayon's invitation to appear on a program about the signing of the Magna Carta in 1215.[84]

The contrast between Murrow and the main voice of *CBS Is There*, John Daly, is instructive. According to the cultural historian Susan J. Douglas, Murrow and his fellow radio journalists during the war had come to embody a brand of "middle-class, American masculinity" characterized by unflappable cool under pressure. That was consistent with the 1939 dictate of the CBS News executive Ed Klauber that an "unexcited demeanor at the microphone should be maintained at all times." Most of the reporters on *CBS Is There* (some of whom Murrow had recruited to the network, such as Richard C. Hottelet) were of the same mold in bringing what *Variety* approvingly called "an utterly dead-pan technique of reportage" to the series.[85]

Daly was different. He held a substantial journalistic background, including a stint as CBS White House correspondent, but he also had dramatic aspirations, going on to play the editor Walter Burns in a 1949 television-series version of *The Front Page* before achieving his greatest fame as host of the TV quiz show *What's My Line?*[86] His persona on *CBS Is There* could be stoical and calm when necessary, but when the historical circumstances of a particular episode dictated it, he could ratchet himself up to a pitch equal to that of a sportscaster, Herbert Morrison reporting the burning of the Hindenburg, or the doomed announcer describing the final Martian advance in radio's *The*

War of the Worlds. That annoyed at least one reviewer, who heard in Daly "a frenzy of excitement of which no modern reporter would be guilty." Still, it did serve the dramatic purposes of the series, whereas Douglas Edwards in the same role could sound bland and disinterested in comparison.[87] It also coincided with the showmanship instincts of Shayon, whose background had been primarily in radio drama as opposed to news.

A prime example was an episode originally broadcast in April 1948. "The Battle of Plassey" reenacted Robert Clive's 1757 military victory that helped deliver India to the British Empire. Daly delivered an "on scene" report describing Clive's troops holding off the climactic enemy charge against them:

> The elephants are still a hundred or so yards away. . . . [Now] they're almost on top of us! . . . We may have to get out of here, and fast! . . . [Musket fire, shouting] The smoke is pouring out of them again, the muskets—the smoke is so heavy, it's difficult to see—it's so thick now that you could almost cut it. . . . [Elephants trumpeting] We don't know what's happening; everything is confused and uncertain until this smoke begins to pull away from here. I for one don't know whether to stand here or to cut and run! . . . [More elephants] We can hear the trumpeting of the elephants; you may be able to hear it in the background. It sounds like a bunch of mad monkeys having a fight—*all the monkeys in the world* in one basket fighting with each other! . . .
> [More shouting, shooting, elephants, etc.] *The elephants have been stopped!* They're no longer charging forward! They're stopped in their tracks and their mounts are hanging on for dear life itself; they can't control them any longer at all! Some of these younger bull elephants are bucking almost like horses [and] rearing and churning the air and screaming! . . . The elephants are completely out of control, completely out of control. They're running wildly away from this musket fire. They're shaking their riders from their backs. The leaders are already charging back into their own camp! . . . *Colonel Clive,* sir—you are now the ruler of Bengal! . . . It's a magnificent victory![88]

Although the program aired just after India had gained its independence from Britain, the show did not discuss colonialism or imperialism (with Clive assuring Daly that "the British are here to trade, not to rule").[89] Still, Shayon said that he made it a point to look for subjects "with some meaning for our time" to help listeners "see more clearly the relationship between the past and the present." Later he would say he sought to infuse the series with "political bite," although some shows showed more fang than others.[90] A program on the signing of the Declaration of Independence recounted the clash between John Adams and John Dickinson over breaking with Britain, but it ultimately adopted the proudly patriotic tone that also characterized much of NBC's

Living 1948. Civil War episodes stressed the theme of postbellum reconciliation hinted at in the Lincoln show. A dramatization of the 1865 surrender at Appomattox included Quincy Howe's analysis of the negotiations: "Only honorable terms can soothe the open wounds of the proud South. Only with honorable terms can Lee return to his men down here in the valley and say [to them], 'We must forget the past.'"[91]

Other programs presented similarly conventional commemorations of American history but spoke to contemporary concerns, if only indirectly. "The Burr-Hamilton Duel," set in 1804, took Alexander Hamilton to task for his low opinion of the common man and woman venerated by 1940s radio writers such as Shayon. "Hamilton has never quite understood the American spirit—our town meeting form of democracy," said Howe. "The Defense of the Alamo" aired when debate was beginning over whether America should support the Marshall Plan for aiding Europe. CBS's Ken Roberts told listeners that although America was officially neutral in the 1836 Texas conflict, Texas had "long been culturally and physically allied to the United States. These people, these Texans, are the cousins and the brothers and the husbands of our own families. They are our kin."[92] Thus American support for their cause—and implicitly for those of our other allies, past and present—was right and natural.

Some episodes more forthrightly expressed Shayon's liberal sensibilities. "The Dreyfus Case," set in 1899, made no mention of anti-Semitism, but it noted worldwide condemnation of France's wrongful imprisonment of Alfred Dreyfus and ended by proclaiming that "the fight against prejudice goes on." The Toussaint L'ouverture show about the 1802 Haitian Revolution against Napoleon also addressed prejudice while once again emphasizing the need for amity. Toussaint quieted a crowd chanting "death to whites" by saying, "I am too much a believer in the rights of man to think that there is one color in nature superior to another." One of Napoleon's former soldiers who had switched to fight under Toussaint said the same: "This is not a war of color, but a revolution for liberty and equality and fraternity for the white man and the Negro." A program on Sitting Bull's surrender in 1881 called the incident "one of the most shameful moments in American history" and featured the French Canadian trader Jean-Louis Légaré taunting a U.S. army officer over Sitting Bull's defeat of Custer: "It is ironic, is it not monsieur, that your Custer was cut down as a tyrant in 1876, the one hundredth anniversary of your own war of independence? . . . [Is your independence] for the red man too, or for the white man only?"[93]

Of particular interest to Shayon was free thought in the face of fear and superstition. "Columbus Discovers America" recalled the Battle of Plassey episode in that it did not address colonialism at the same time that it patronized the mysterious people whom the Spanish expedition met in the New World in 1492. ("Seems that they're painted every color of the rainbow, but otherwise they haven't got a stitch on!" exclaimed Daly in his role as observer aboard the expedition.) For Shayon, though, the point was that Columbus "represented a triumph of imagination over traditional ways of thinking."[94] In a like spirit, *CBS Is There* decried the 1692 Salem witch trials ("If this is contempt of this court, let it be so!" Daly said in criticizing the conviction of one supposed witch), and the series characterized the Puritans' 1637 banishment of free-minded Anne Hutchinson (whom it called "the first American feminist") as "an obscure but moving episode in the fight for freedom of speech and conscience."[95]

Perhaps the most notable example was "The Death of Socrates," which took listeners "back almost 2,347 years to that fateful day [in 399 B.C.] when one of the most enlightened democracies on earth trembled on the brink of a cup of poison." Plato was heard saying of his fellow Greeks that "we are confused, desperate. And so we seek someone to blame and sacrifice." He attributed the animus toward Socrates to those who clung to "old ideas that fear to be challenged, cross-examined, and perhaps exposed as no longer virtuous in our time," adding that if Socrates were to die, "it will not be democracy that's to blame, but those who have lost faith in democracy." Just before drinking the fatal hemlock, Socrates himself (played by Walter Hampden) condemns his judges, saying that if they believe that they could "restrain men from reproaching them for the evil of their lives, they are very much mistaken. It is much better for you and much easier not to silence reproaches, but to make yourselves as perfect as you can." The program originally aired in March 1948, a few months after the HUAC hearings on alleged communism in the movie industry. Reviewing the show in the *New York Herald Tribune*, John Crosby observed that its dialogue "might easily be applied to the hysteria behind our current witch hunts."[96]

Critical support for the series remained high, and it won a Peabody Award. However, despite the initial interest from potential commercial sponsors that *Variety* had reported back in the spring of 1947, CBS failed in repeated attempts to attract such a sponsor. In a promotional brochure, the network pointed to listener calls for the series' renewal in claiming there was "every reason to expect *CBS Is There* to win not only *loyal*, but *large*, audiences under sponsorship."[97] In May 1948, the series was renamed *You Are There*, the thought being that removing the network name from the title might make it more attractive to sponsors not wishing to share credit for the series.[98] Rumors of American Oil Institute sponsorship proved unfounded, but Shayon was unconcerned. In the interests of quality, he spent anywhere from $2,000 to $2,900 of CBS's money per episode, which he proudly told a critic was "a high price for a sustaining program." Top actors clamored to appear on the series. As Shayon recalled in his autobiography, his time producing *CBS Is There* constituted "my happiest years at CBS."[99]

* * *

Shayon's happiest years coincided with the high-water mark of postwar audio documentary at the networks. Just as CBS had moved aggressively into documentary production to gain a competitive edge over its rivals, so NBC created *Living 1948* and similar programs to put it "into the running for the industry kudos it has seen passed out to other nets in the past for public service programming of this format," as *Variety* commented at the time. Robert Saudek and ABC did the same to gain distinction for the fledgling network. Following the airing of *Communism—U.S. Brand*, Jack Gould wrote that in "vision and understanding of radio's educational and informational potential, Mr. Saudek's department is now well out in front."[100]

The forms of the documentaries ran the gamut from the straight docudrama of the ABC communism program, docudrama blended with polling data in *Living 1948*, to docudrama combined with recorded actuality in *V.D.: The Conspiracy of Silence*. Alongside those were the straight actuality format of *Slums*, the musicalized *1960?? Jiminy Cricket!*, and the history-cum-reportage of *CBS Is There*. The rich experimentation took advantage of what Susan J. Douglas has called the "dimensional listening" that radio can offer, which she compares with "informational listening." The latter is "a relatively flat kind" of experience in which audiences absorb facts but little else. Dimensional listening appeals to the imagination, inviting listeners to picture themselves inside a present-day tenement or outside Ford's Theater in 1865 or aboard a rocket to the future with Donald and Jiminy and Grumpy. "This listening is work," says Douglas, "but it is also highly gratifying because it is your own invention."[101]

Wartime radio news as exemplified by Edward R. Murrow's dramatic and heroic-sounding reports from London had exploited radio's capacity for dimensional listening in a way that underscored the power of broadcast journalism. The same was true of postwar documentary, including—or especially—*CBS Is There*. Far from undermining newscasters' credibility as Murrow had feared, the series enhanced it. "News on radio in the 1940s carried an unusual authority over the public mind," Shayon wrote in his memoirs. He recalled the visceral response of listeners to his historical series, including some who were convinced by the episode "The Last Days of Pompeii," set in 79 A.D., that modern-day Italy was being buried in ash and molten lava (suggesting that perhaps CBS executives had been right to worry about their audience's credulity). "The phenomenon was possible only in radio, where the listener's mind sets the stage," said Shayon. "It would never happen in television, where the visual image is a reality check."[102] Add to that the series' conceit that CBS was not just "there" but everywhere in time and space across the ages. "'CBS Is There' exhibits a news sense the like of which probably never before has been seen in the annals of news gathering—and quite possibly never will be again," the *Christian Science Monitor* observed at the time. Likewise, an anonymous CBS staffer jokingly told *Time* that the "series need end only with the crack of doom—recorded and transcribed by CBS, of course."[103]

The documentaries also encouraged listeners to imagine a better home that they themselves could help make, drawing upon history's lessons. The programs embodied much of the same liberal optimism that had carried over from World War II docudrama. They declared that schools and hospitals could be improved, slums could be eliminated, disease could be eradicated, and freedom of thought could endure. In the spirit of Socrates's speech on *CBS Is There,* they suggested that reproaching the world for its shortcomings would help it to perfect itself. There were limitations to their worldview; as Michele Hilmes wrote of the wartime shows, they "combined a sentimental appeal to basic American values with a not-so-subtle threat that the whole world [was] watching."[104] Likewise, they typically expressed a white, male, middle-class perspective. Women and minorities were sometimes sympathetically depicted, but they were just as often caricatured, as with Kitty of the "shabby past" in *V.D.: The Conspiracy of Silence* or the natives encountered by Columbus's expedition in *CBS Is There.* When the poor were allowed to speak for themselves in *Slums,* they were cast as largely helpless victims. The vast, undifferentiated "America" to which *Living 1948* sent greetings each week accompanied by the latest Gallup numbers implied a mass population to be addressed by elites rather than the more truly participatory democracy conjured by programs such as *The Eagle's Brood.*

The liberal consensus that the programs addressed was beginning to crack by 1948, as Norman Corwin had discovered at CBS. Arnold Perl found that out for himself while writing a documentary for Mutual, *To Secure These Rights.* Originally it was intended to be a four-part dramatization of a report from President Truman's Committee on Civil Rights. Perl produced a script that reportedly "spoke up frankly and fearlessly against lynching practices in the South." The network president Edgar Kobak vetoed the script, preferring to let listeners "make up their own minds" about the report's findings.[105] Critics saw it as a move to appease Mutual's southern affiliates that otherwise might refuse to carry the program. *To Secure These Rights* finally aired as an unadorned reading of the civil-rights report, similar to what ABC had done with John Hersey's *Hiroshima* a couple of years previously. As it turned out, a higher percentage of Mutual's stations in the South carried the program than did its stations elsewhere in the country. Still, fewer than half of its affiliates nationwide both carried and promoted the show, and Perl accused the network of censorship, seeing the rejection of his original script as part of a broader media clampdown on liberal writers and viewpoints.[106]

The controversy concerning *Communism—U.S. Brand* was further evidence of the shifts under way. Norman Rosten was a One World liberal in the mold of Perl and Corwin, whereas Morton Wishengrad was a liberal unalterably opposed to communism and deeply suspicious of the Soviet Union. Although Rosten predicted to Wishengrad that "the rewards of the communist-hunting liberal will be bitter and brief," the opposite proved to be true. The cold-war tensions that underlay *Communism—U.S. Brand* (and that *Living 1948* more temperately addressed) empowered anticommunism, as signified by the enthusiastic response of the liberal-minded Jack Gould to Wishengrad's program. Sentiments favoring One World and rapprochement with Russia fell into disfavor, with Rosten, Perl, and Corwin eventually paying a heavy price.

In addition, the competitive pressures that originally had enabled documentary began to work against it. Not only was CBS increasingly devoted to surpassing NBC in the ratings, but also television was starting to take center stage. *Living 1948* had gently mocked the politicians preening for the TV cameras at that year's political conventions, and in a program titled "Television: Many Happy Returns?" it had predicted great things for the new medium: "To charm the air / To give to all—rich and poor alike— / Seats at the elbow of the statesmen; / Presence at the pregnant events of the nation."[107] Radio executives were less sanguine. CBS's Hubbell Robinson summarized the viewpoints of many: "Television is about to do to radio what the Sioux did to Custer. There is going to be a massacre." Robinson thought that such fears were exaggerated, but he also said that for radio to survive, it would have to innovate with fewer resources than in the past: "It is not a simple question of money; in fact, showmanship, ingenuity, and resourcefulness are going to have to do what many producers have attempted to achieve by piling up dollars in production costs—a procedure that is doomed to failure."[108]

Against that backdrop, Robert Lewis Shayon's loyal crew rang out 1948 with a surprise New Year's Eve gift—a mock program titled "Here You Are" that they wrote and recorded without his knowledge. The actors who appeared regularly on the series impersonated and satirized Shayon, CBS's journalists, and themselves, claiming to base their production on "snide remarks, overheard comments, and historical misrepresentation." It began with a dour "Don Hollenbeck" (*"goddammit,* get off the mic cable!") in the midst of a milling mob of radio performers who were outside the "heavily barred and guarded door" of Shayon's office and desperately seeking to be cast in his series. "*What* Mr. Shayon is about to cast, no one knows," said "Quincy Howe" in presenting his usual expert analysis. "Rumors have been flying thick and fast. Prevalent among them is the theory that his next production may be either 'The Discovery of Water—Sixty-Three Years after Dirt,' or 'The Turning of the First Wheel, and How Surprised Men Were.'" Concerning the state of radio, "Richard C. Hottelet" offered his take on "what this all means for a rapidly dying industry."[109]

As for the ubiquitous "John Daly," he was stashed with a portable transmitter inside a file cabinet in Shayon's office, where the only sound was the "swish of darts" flying through the air. Daly was overcome with emotion when he related the fate of the notoriously gaunt Hollenbeck at the hands of the mob outside—the actors had mistook him for a microphone and "plugged him in," with fatal results. The parody ended with the usual echo-slathered announcer's voice: "December 31, 1948: Another actor leaves, and Shayon continues to cast." It was followed by a tune sung by a quartet recruited from Arthur Godfrey's daily CBS show:

> When CBS is there, you are there
> Two-thirty Sunday is when we are on the air
> So listen while Bob Shayon tears his hair
> You are there, you are there.

The crew presented Shayon and his assistant Lorraine Doherty with the recording of the production and an attached note: "FROM THE CAST OF 'YOU ARE THERE,' 'HERE YOU ARE': . . . [W]e the below named cast of 'Here You Are' accept full responsibility for hoping that this is the *Happiest New Year* you both have ever known."

Alas for Shayon and his assistant, it was not to be.

5

The Quick and the Dead

"Is the documentary dead?" Addressing that question in *Variety* in July 1949, the CBS vice president Davidson Taylor declared that it would survive "so long as Americans refuse to believe that the state knows all the answers, and are convinced that the people must decide things for themselves." But he acknowledged that audio documentary was facing new challenges at a time "when money is scarce and radio is undergoing a technological revolution." Prominent documentarians such as Robert Heller had moved into entertainment programming. Heller suggested that audiences might learn more about the plight of underpaid teachers from a comedy series such as *Our Miss Brooks* than from a public-affairs program.[1]

Other documentarians were leaving commercial broadcasting altogether, not necessarily voluntarily. Following his conversation with William Paley, during which the CBS chief had told him to write for a broader audience, Norman Corwin received a proposed new contract whose terms seemed calculated to ensure that he would reject them. He left the network in early 1949 for a new position producing radio shows for the United Nations. One of those shows aired on CBS the following summer, with Corwin in effect working as a freelancer in cooperation with the CBS Documentary Unit. *Citizen of the World* reiterated the themes of *One World Flight* while using dramatizations and sound effects in addition to the type of actuality recordings that the earlier series had introduced. The program acknowledged the problems confronting the globe, including the "Big Five" of hunger, erosion, disease, poverty, and war, but it ended on a typically hopeful note via the narrator, Lee J. Cobb:

When long years from now the full returns are in and there can be a grand accounting, perhaps someone will place a footnote with a bright star marking it: "Around mid-century came the first Citizen of the World to gang up against the terrible Big Five, to take them on." Meanwhile, tomorrow is tomorrow, and the year 2000 is a long way off. There's work ahead, rubble to clean up, quarrels to settle, fears to quell, mouths to feed, a thousand chores to do. And in the doing, the Citizen of the World will be right there pitching with the rest of us who care. For to him and what concerns him, a seed is not too little, nor is all humankind too big.[2]

Corwin at least was able to leave CBS on his own terms; Robert Lewis Shayon was not so lucky. Not long before Davidson Taylor wrote in *Variety* of radio's money woes, he called Shayon into his office. "Bill Paley has started a cost-cutting campaign," Shayon recalled Taylor telling him. "You have two weeks in which to leave." Shayon's assistant Lorraine Doherty and *You Are There*'s top two writers also were fired, along with almost two hundred others at the network. The stated rationale was that CBS needed to improve efficiency in the face of increased competition and reduced advertising revenue.[3] As *Variety* had reported earlier in the year, though, the real "heavy" was "television with its grinding demands for more coin and fresh calls for red ink," leaving sustaining radio series such as *You Are There* particularly vulnerable. By cutting that show's chief personnel, CBS saved more than one hundred thousand dollars that it could spend elsewhere.[4]

Along with the economic pressures and the move toward television, documentary faced another technological change: the introduction of plastic audiotape, which as a recording medium proved vastly superior to wire. NBC had finally dropped its ban on airing recordings at the start of 1949, following CBS's lifting of its own ban a couple of months previously. NBC had acted primarily to fend off additional talent raids by CBS (following Bing Crosby's lead, radio stars had come to prefer recording their shows as opposed to having to perform them live).[5] However, the network news head William Brooks said that there were other benefits: "Tape recording units have proven so flexible that we are constantly finding new uses for them in news reporting."[6] That summer, NBC debuted a weekly series centered around recordings, *Voices and Events,* with James Fleming billed as "editor-in-chief." The network said that the program aimed to present "a balanced diet of the week's news, permitting the tape recordings to tell the story as completely as they can, filling in with narration only when necessary to advance the sense of the story or to make a transition."[7] Whatever fresh possibilities such a format offered to news and documentary, they also pointed toward a diminished role for those such as Corwin and Shayon, whose craft was built around writing narration and dialogue.

Underlying all else were the shifts within the American liberal spirit that had sparked so much of wartime and postwar documentary. As one historian has described it, a liberalism that had "stressed the goodness and rationality of man and the possibilities for improving society" was giving way to one that "combined a striving for social reform with a certainty of human frailty and a suspicion that the ideal society was but a naïve dream."[8] That was demonstrated by the conspicuous failure of Henry Wallace's 1948 presidential campaign and the simultaneous rise in influence of the centrist Americans for Democratic Action (ADA). Wallace had been backed by the Progressive Citizens of America (PCA), with Shayon and Corwin both speaking at PCA-sponsored events and Corwin enthusiastically supporting Wallace. In contrast, the ADA scorned Wallace and his backers while assuming a tough posture toward the Soviet Union and communism.[9]

The ADA's cofounder, Arthur Schlesinger Jr., presented what has been called "the classic manifesto of liberal anticommunism" in his 1949 book *The Vital Center*.[10] Schlesinger defined communism and fascism as flip sides of the same totalitarian coin. Much of his argument paralleled that of Morton Wishengrad's *Communism—U.S. Brand*: social reforms were necessary, but not with communist support, because "the personal word of the Communist is worthless and cooperation with him impossible." Like the hapless Phil Blake in the radio program, the typical communist was "lonely and frustrated" and thus easy prey to party manipulation. That said, American communism's real threat was not toward overthrowing capitalism but toward "dividing and neutralizing the left," particularly the "Doughface progressives" with their "pervading belief in human perfectibility which has disarmed progressivism in too many of its encounters with actuality." One needed only to look at those writers who had "exerted their influence toward lowering and softening artistic standards in a pseudo-democratic direction," as exemplified by the "phony populism" served up by "the incredible radio plays of Norman Corwin."[11]

Schlesinger's contempt for Corwin was returned in kind, with Corwin privately expressing his "feeling that gradually all of the artists and intellectuals will move out of the country and leave it all to Arthur Schlesinger, Jr., Arthur Godfrey, and Coca Cola."[12] Regardless, the backlash against Corwin that had begun with *One World Flight* intensified. In August 1949, U.S. Senator Pat McCarran (who later would chair a Senate investigation of alleged subversion in broadcasting) charged that the writer had been affiliated with numerous organizations cited as "Communist and subversive," although

McCarran denied press reports that he had branded Corwin himself as a communist. Corwin retorted that he was "a vastly better patriot" than the senator, adding that denying charges of communist sympathies was "getting to be a daily exercise for anybody who ever entertained a liberal idea."[13] That state of affairs would only grow worse for him, as it also would for Shayon. In the meantime, radio documentary reflected the changed political climate via a new group of practitioners, among them a pacifist chaplain.

Elmore McKee and *The People Act*

In a written proposal for what would become the NBC series *The People Act*, Elmore McKee invoked both Arthur Schlesinger Jr. and another well-known liberal anticommunist, Sidney Hook. McKee wrote of "the need for a clearly defined democratic counter-offensive" against totalitarianism, borrowing language from *The Vital Center*. The idea was to create a radio program "in support of Mr. Schlesinger's plea that we show that there do exist in our land vital voluntary associations in which the individual derives a real sense of belonging, and in which the schism between individual and community is healed." Likewise, it would embody "Mr. Hook's suggestion of the need for widespread study of democracy as a way of life" at a time when Hook said that the nation's colleges were threatened by communist-tainted thought.[14] In brief, the series would promote the sort of grassroots citizen action that Shayon's *The Eagle's Brood* had espoused a couple of years previously. This time the implicit aim would be toward not only solving community problems but also winning the cold war.

The notion of McKee fighting any kind of war seemed improbable. A 1919 Yale graduate, he had served as campus chaplain there and then as a pastor in Buffalo before becoming rector of St. George's Protestant Episcopal church on East Sixteenth Street in Manhattan in 1936. It was described at the time as "a parish where the rich and poor meet side by side and worship together."[15] Over the next decade at the church, McKee promoted several liberal-minded causes. "If overcrowding threatens the strength of our population, let us advocate the distribution of birth control information; or if the sad conditions of the tenements persist, let us work for model housing," he preached soon after assuming his post at St. George's. He condemned Nazism and anti-Semitism following the Kristallnacht pogrom in Germany in 1938, and the following year he called the Daughters of the American Revolution's banning of a performance of the African American opera singer Marian Anderson "quite pagan."[16]

More controversial was McKee's forthright pacifism. Just before America's entry into World War II, he declared that "war is an even greater enemy than Hitler" at a meeting of antiwar Episcopals. (Two New York City churches had refused to host the meeting before a site for it was finally found.) McKee also opposed the training of air-ward wardens and other defensive preparations prior to the war. Shortly after Pearl Harbor, he wrote to the New York Times to protest the derogatory term "Jap," saying, "Feeding our sense of superiority toward oriental peoples is a good way to lose the next peace as we did the last one." Two years after that, he joined other pastors in urging a halt to air raids against German cities, calling them a "carnival of death."[17] According to a senior warden of St. George's church, such views differed from "the vast majority of our members." Still, the warden insisted that that had nothing to do with McKee's decision to resign the rector's position in 1946; instead, as McKee later said, he wanted "to do public service of a different sort."[18]

Late that same year, McKee traveled to devastated Germany to help create self-help neighborhood centers. There he had an epiphany. He discovered that the Germans had to be taught not to rely on experts or other authority figures in rebuilding their country. "I felt I was confronted with some of the social ingredients that could tip the balance for the survival of civilization," McKee recalled. He saw anew that "democracy means that everyone concerned with a problem should have something to say about it and step forward to take responsibility." Moreover, he came to believe that this was not happening nearly enough in America. McKee returned home eager to help foster "a social climate of faith in people and of expectation that self-determination and a wide distribution of responsibility would work."[19] Such a viewpoint was consistent with Arthur Schlesinger Jr.'s stated faith in "the integrity of the individual" who was "most humane in a society which distributes power widely"; it also was consistent with the pluralist political philosophy that many liberal intellectuals (Schlesinger and Sidney Hook among them) embraced after the war as an alternative to Marxism. Finally, it extended a belief that McKee had expressed back in the 1930s: that America had to reject the mob psychology so easily exploited by demagogues and remember that one of its "greatest contributions to modern civilization is its ability to get people to work together cooperatively."[20]

In the fall of 1949, McKee approached NBC with his idea for a radio series that would send him across America to document specific examples of cooperative democracy in action. The network in turn approached the Twentieth Century Fund to underwrite the program; as one NBC official put it, such organizations served as "financial angels" in lieu of commercial sponsorship.[21] The Twentieth Century Fund was considered a liberal-leaning think tank, although its board of trustees included Charles Taft, the brother of the staunchly conservative Republican Senator Robert Taft. Fund officials agreed to spend an estimated $1,500 a week for a thirteen-week series. NBC retained editorial control and provided the scripts and narration under the auspices of its ongoing *Living* series. The network originally proposed broadcasting it on Saturday afternoons during the summer of 1950, but the fund balked, believing that audiences would be too small then. Instead NBC waited until the end of football season (during which it broadcast games) to air the series on the weekends starting in December and continuing until March 1951.[22]

Early in the preparation of what eventually was titled *The People Act*, McKee recorded a test program that an NBC production director deemed unairable. He said that McKee was "not a radio professional" and that in his commentary segments "the show simply drops dead."[23] Instead, the longtime *Living* host Ben Grauer would handle the narration with only brief introductions and summations from McKee, and Lou Hazam would do the brunt of the writing. Hazam accompanied McKee on his journeys across the country, which according to McKee made for much stronger scripts. The two men also took advantage of NBC's recent investment in recording technology, at times using a new Minitape machine to do field interviews.[24] In a report in the midst of his travels, McKee wrote that the radio programs would highlight "bulwarks against exaggerated centralism in government. These are mainstreams of democracy. Hitler would never have tolerated such autonomous cells of vitality as we shall watch in action in this series."[25]

The People Act debuted on December 9, 1950, with the episode "Miracle on the Mount." It billed itself as "the first of a special series of programs designed to penetrate America to its grassroots and discover the characteristic and democratic ways in which American citizens unite to solve some of their everyday problems."[26] Ben Grauer delivered his customary salutation ("Greetings, America!") before relating the story of a country doctor in Bat Cave, North Carolina, who had vainly struggled for years to improve the medical facilities of his rural district. "Neither the state nor the Federal Government will give us money to help build a small hospital like we need," Grauer quoted the doctor as telling his fellow citizens. "Can't we do it ourselves? Can't we build *our own* answer?" Indeed they could, raising the funds with the help of the local Elks Club to buy a dilapidated schoolhouse and convert it into a modern hospital. The program included actualities from the doctor and his neighbors. "What you hear today are not actors performing for you," said McKee in a short introduction, "but the flavorful voices of the people of Hickory Nut Valley themselves just as I recorded them."

At the end of the program, Grauer engaged in scripted conversation with the executive director of the Twentieth Century Fund, Evans Clark. He told Grauer that a lack of adequate medical care resulted in the equivalent of removing "a million and a half able people from our working force and keeping them idle for an *entire year*." Grauer called it "a shocking waste—especially when the nation is straining its resources for a rearmament program." (The United States was back at war in Korea, and as of December 1950, the battle was not going well.) Clark responded by citing the fund's report, *America's Needs and Resources*, which had inspired ABC's Jiminy Cricket documentary three years previously. It said that with only a modest additional investment, the country "could provide proper medical care for our entire population" by 1960.

Succeeding programs followed the same format—Grauer relating an inspirational real-life tale of citizen action via a varying blend of music, recordings of the actual participants, and sound effects or dramatic recreations. The program always concluded with a Twentieth Century Fund representative discussing with Grauer or an NBC journalist the implications of the story just heard. The programs were recorded and edited in the studio before being broadcast, a significant departure from the recording-ban era, when most documentaries were produced live.

Two episodes about labor relations demonstrated the different techniques and common themes that the series offered. "Partners in Velvet" reported on the American Velvet Company in Connecticut, where the son of the late, fearsome owner reversed years of labor-management strife by offering the workers a profit-sharing plan. The program featured recordings of the son and the workers but told its story mostly through dramatized reenactments of highlights in the company's labor history. In contrast, "Cloth of Many Colors" presented its account of New York City's garment workers almost entirely through recorded actualities. The program celebrated the industry's practice of bringing together elected representatives of labor and management with an impartial mediator to resolve disputes. According to a member of the Twentieth Century Fund's labor committee speaking in "Partners in Velvet," such programs exemplified "what we should all strive for in our economic relationships—the respect of human beings for one another—the willingness to see each other's point of view and work together in a democratic way."[27]

The People Act reported more often on small-town America than on big cities such as New York. Many episodes followed the lead of "Miracle on the Mount" in focusing on the rural South. "The Sylvania Story" told of an Arkansas town where "halfway through the twentieth century, the people have turned pioneer again, and in so doing have not only revolutionized their own living conditions, but have demonstrated the vitality of democracy." Depression and erosion had given way to modern crop and soil management, with a helping hand from the "C. C. boys" (the Civilian Conservation Corps of the New Deal era). "The City That Refused to Die" reported on the resurrection of Decatur, Alabama. The townsfolk had diversified their economy beyond cotton production and built a new dairy and poultry plant. Again, government had helped through the Tennessee Valley Authority, but McKee stressed in the program that the town had achieved its greatest success when it "decided to quit asking for help from outside—decided to develop to the full its own resources." "Red Clay and Teamwork" related a similar story from Carroll County, Georgia; the citizens had formed a countywide service council that had helped build new schools, libraries, and hospitals as well as a church. The Twentieth Century Fund's Evans Clark said that the new church was especially symbolic, given that other countries routinely saw "clergy thrown into prison, churches suppressed, and religion under the heel of government dictatorship."[28]

"Red Clay and Teamwork" referred briefly to an interracial committee that the service council had formed, but otherwise the southern-themed stories did not allude to African Americans. When the series did give voice to people of color, they tended to be portrayed as the beneficiaries of white munificence. "Crusade in Baltimore" recalled ABC's *Slums* in presenting recordings of slum residents, including a woman recalling how rats had bitten two of her children. The lot of "these unfortunates" was being improved thanks to a "young, attractive, forceful" Smith College student whose research project had helped document the wretched conditions and to a crusading editor who had reported the situation in his newspaper. A new housing division was formed as part of what the mayor called an all-out attack on slums. Likewise, in "The Women Did It," the people of Lawton, Oklahoma, banded together to annex the impoverished neighboring community of Lawton View. As the episode's title implied, women had led the initiative, including one who recalled her initial ignorance of conditions in Lawton View: "We had people from there working in our homes, but I never actually looked over the places where they lived very carefully before." The Twentieth Century Fund's Charles Taft praised Lawton for helping to "keep our democracy alive and real and form the best possible foundation for an offensive on totalitarians that would destroy it." He added that he was not surprised at women's prominent role, calling them more practical than men and less prone to give up easily.[29]

Strong women were featured regularly on *The People Act*. In "The Prairie Noel," Morganville, Kansas's "most world-minded citizen," Velma Carson, prompted her town to adopt Feves, France, as a sort of sister city and send powdered milk, diapers, tomato seeds, and other provisions to a people still suffering mightily from the war's aftereffects. ("Before the war came, we tried isolation," Carson said. "This time we thought that surely we could reach out to the world instead of waiting until they came to us.") "The Sun Shines Bright" described how Kentucky schoolteachers led an effort to improve not only the schools themselves but also their water supply, the roads leading to them, and the diets of the pupils who attended them. Similarly, in "As the Children Go," the women of Haddon Township, New Jersey, helped overcome a failed referendum and opposition from the local taxpayers association to garner enough support to build a new school, thus redressing severe classroom overcrowding. "I don't think we'll ever go back to being just normal housewives again," said one.[30]

Obstacles successfully overcome was a constant theme. "Home Is What They Made It" (the title apparently borrowed from Lou Hazam's old series) told of how a group of returned war veterans and their wives cooperated to build new houses of their own in Lorain, Ohio. They managed to keep going even after money for the project ran out multiple times. "I was a rifleman in the infantry in Italy," said one veteran. "I slogged through a lot of mud there and I slogged through a lot of mud here, but they both were worth it in the end." "Our Partner—the Public" ended the series' original run by again touching on labor relations, this time in the strike-and-strife-ridden city of Toledo, Ohio. A Labor Management Citizens Committee managed to maintain order even through sometimes contentious negotiations over wages and working conditions. "Each week we have brought you a true story with real voices and real people of Americans in vigorous action to solve their own economic and social problems," Elmore McKee said at the program's conclusion. "Every program has proven the truth of what Mrs. Frances Spaeth said, one of those who fought hard for the new school in Haddon Township, New Jersey: 'This struggle has taught us the responsibility that goes with freedom.'"[31]

The People Act won praise, with *Variety*'s reviewer saying that it "was extremely effective" and that it "smacked of authenticity." However, the reviewer did raise the old concern of scheduling: "Saturday night is not one in which the American public is happy to hear about problems. It would gain a wider and more [appreciative audience] in some other time segment."[32] A few years later, Elmore McKee raised a couple of his own caveats in a book based upon the series. Although expressing continued enthusiasm for the project, McKee added that "these stories make no case for over-confidence. Frankly, they were hard to find." In at least one instance, good feelings had darkened, if only temporarily. According to McKee, the people of Morganville, Kansas, had for a time begun questioning their relationship with their sister village in France, which was then "teetering economically, and swinging far to the left of what seemed to them the respectable Left." Morganville had thus fallen prey to the "conflict, confusion, suspicion," and "resurgent isolationism that swept America"—by-products of the same cold-war tensions that had inspired *The People Act*'s "democratic counter-offensive" in the first place.[33]

Fred Friendly and *The Quick and the Dead*

The cold war and anticommunism permeated much of radio programming in 1949 and 1950 after the Soviets detonated their first atomic bomb, the communists took control of China, and the Korean War erupted. *Variety* reported that broadcasters "out to hypo their ratings" were "tailoring scripts to have the villains directly labeled as Commies . . . characterizing the Reds as speaking with heavy accents and practically recreating the old concept of the bearded Bolshevik carrying a bomb."[34] NBC's *Living* series reflected the same trend, even if it did not consciously seek to inflate ratings or traffic in melodrama. In an October 1949 script by Lou Hazam, the series warned that "if we do not counteract Russia's wooing" of the people of Southeast Asia, "we may well lose the battle by default." It also declared that South Korea "cannot fully realize full economic recovery constantly under the shadow of the Soviet-dominated North," adding that the South would be hard-pressed to resist an invasion from the North."[35]

After North Korea did in fact invade South Korea, *Living* looked toward "*other potential Koreas,*" including Vietnam. It called Ho Chi Minh a "trained mouthpiece" for Stalin and said that Indochina represented "an American frontier with Russia." That in turn demanded "*strength—America's mobilized strength!*"[36] In a similar vein, under the executive producer Wade Arnold, the series presented a five-part report titled "Malice in Wonderland" that was designed to expose Soviet propaganda. The debut was scheduled for Labor Day weekend, with NBC saying that the opening segment would "contrast the promises of Marx and Lenin with the facts of life as the average worker lives them behind the [Iron] [C]urtain."[37] Subsequent programs in the series quoted Charles Siepmann, the onetime Blue Book consultant: "The true crisis of our times is the struggle for the allegiance of men's minds. The battleground is the minds of men, the extent of the battleground is global. *The chief weapon used is propaganda.*" Ben Grauer's narration elaborated: "The [Soviet] bullets are words, ideas, slogans. They are aimed to discredit, disarm, divide. For you [listeners], the hero and heroine of this great suspense story of our time, the end of this process is slavery and death." America had to fight back with its own words and ideas: "We have the greatest story in the world to tell—the story of freedom and truth—and we must tell our story now, and tell it well, to all the world."[38]

In its own way, *The Quick and the Dead* represented another NBC effort to promote a uniquely American version of freedom and truth during the cold war, although it did not issue nearly so blatant a call to arms as that proffered by "Malice in Wonderland." Instead, the series explored the origins of the most terrifying arms of all, atomic and nuclear weapons. Its author was another relative newcomer to network radio documentary, if not to radio itself.

Fred Friendly's broadcasting career dated back to 1937 at WEAN in Providence, Rhode Island, where he had spent most of his formative years. He produced a history program for WEAN called *Footsteps in the Sands of Time*. During the war he discovered the wire recorder while working as an army information officer. At the time, he compared the machine to "'a magic carpet . . . taking a man with a story to Budapest, to the Queen Mary, to the Taj Mahal, and by golly, maybe even the moon and the stars in our lifetime.'"[39] He also flew in a reconnaissance plane over Hiroshima and Nagasaki shortly after the bombs were dropped. Friendly later said that that memory, along with those of seeing concentration-camp and Bataan Death March survivors, were "the only scars I bear from what was, for me, a relatively soft war."[40]

The atomic attacks would serve as the climax of Friendly's first collaboration with Edward R. Murrow, *I Can Hear It Now,* which appeared in late 1948. Friendly had wanted to produce a recorded history album in the vein of the radio show he had done before the war. He shared his idea with Murrow's agent, John G. "Jap" Gude, and suggested that Murrow would make an excellent narrator for the project. Gude arranged a meeting between the two men, and Murrow readily agreed to participate.[41] This would not be a dramatized re-creation of history such as *CBS Is There,* toward which Murrow had always been leery (Friendly similarly would tell Robert Lewis Shayon that the series "violated the separation of news from entertainment").[42] *I Can Hear It Now* would exploit the new media of magnetic audiotape and the long-playing record in a straightforward presentation of actual voices and broadcasts from 1933 to 1945. Still, Friendly did not eschew dramatizations or re-creations entirely. John Daly's 1941 announcement of the bombing of Pearl Harbor was doctored to make it sound as though it interrupted the beginning of a live broadcast of the New York Philharmonic. In addition, Friendly asked the newscaster Robert Trout to rerecord his bulletin announcing the 1945 Japanese surrender on the grounds that the original recording was flawed. When Charles de Gaulle refused to reread a 1940 radio address calling for French resistance to German occupation, it enraged Friendly.[43]

No special production tricks were required for the record album's conclusion, which recounted the bombing of Hiroshima. First there was a recording of a prayer read for the Enola Gay crew at their airbase before the mission; then came Murrow's brief, stark description of the bombing and an excerpt from President Truman's announcement of the attack. Murrow in his narration cited the words of William Laurence, the *New York Times* science reporter who had witnessed the first atomic-bomb test in the New Mexico desert. Laurence had compared the explosion to "the dawn of creation," and Murrow wondered aloud if history would prove that to be so: "Was there to be still another cycle of affliction, appeasement, and annihilation? Or had we walked through midnight toward the dawn without knowing it?"[44]

I Can Hear It Now became a surprise hit for Columbia Records, taking advantage of a musicians'-union recording ban and the resulting demand for fresh material. It earned Friendly substantial royalties.[45] Meanwhile, he had begun working at NBC, where he created a radio quiz show that soon moved to television, *Who Said That?* The host Robert Trout was at once fascinated, amused, and appalled by Friendly and maintained a written log of his eccentricities. "He is very changeable, and when he changes he goes frequently from one extreme to the other," wrote Trout.[46] The same temper that Friendly had displayed upon being snubbed by the leader of free France erupted regularly when he was confronted with the challenges of producing a live show at the start of the TV era. A lengthy 1949 memorandum to his superiors bemoaned "the mire of quicksand which threatens to suffocate [*Who Said That?*] and strangle it with lethargy and ineptness."[47] Friendly's coworkers on the show bore the brunt of his outbursts. Trout reported in a memo to NBC's vice president of news, William Brooks, that "Friendly departed in a cloud of dust and gore" after one broadcast. On another occasion, a long-suffering colleague wrote to Brooks, "Friendly blew his top again. Give him a vacation. Or better still I'll take one."[48]

At the start of 1950, Brooks gave Friendly a new assignment that would not subject him to the vicissitudes of live television and that also would take full advantage of the talents he had displayed on *I Can Hear It Now*. President Truman had ordered development of the hydrogen bomb at the end of January while invoking his responsibility "to see to it that our country is able to defend itself against any possible aggressor."[49] That followed the president's announcement the previous September of the Soviets' first atomic-bomb test. Against that backdrop, Brooks directed Friendly to develop a radio documentary on atomic weapons and energy.[50]

The public mood concerning atomic weapons had shifted appreciably since the broadcast of antinuclear programs such as *Operation Crossroads*, *Hiroshima*, and *Unhappy Birthday* soon after the war. According to the historian Paul Boyer, "The dread destroyer of 1945 had become the shield of the Republic by 1950; America must have as many nuclear weapons as possible, and the bigger the better, for the death struggle with communism that lay ahead."[51] William Laurence would become a particular target of critical observers such as Boyer. Originally from Lithuania, Laurence began writing for the *New York Times* in 1930 and made his name as a pioneering science reporter. In May 1945, the government tapped him to tell the story of the development of the atomic bomb. It was a highly secret assignment; not even his wife or editor was told what exactly it involved. Laurence was the only reporter permitted to see the first detonation of the bomb in New Mexico in July and the only one allowed to witness the bombing of Nagasaki the following month. His stories earned him a Pulitzer Prize, along with the moniker "Atomic Bill."[52]

However, in exchange for exclusivity, Laurence was required to write government press releases about the bomb. That included a draft of President Truman's announcement of the Hiroshima attack. Laurence saw it as an extraordinary opportunity and privilege; as he put it, "No greater honor could have come to any newspaperman, or anyone else for that matter."[53] Some observers since then have condemned his reporting as being "severely compromised" and as representing "an emotional celebration of the science, engineering, and industry behind the bomb" while downplaying the negative effects of atomic warfare and the burgeoning arms race.[54]

Nonetheless, it was Laurence to whom Friendly and NBC turned in 1950 to help create *The Quick and the Dead*. Laurence had recounted witnessing the bomb's development in his 1946 book *Dawn over Zero*. NBC was eager to draw upon "the only reporter allowed backstage when the atomic project was undertaken," and it billed *The Quick and the Dead* as "mainly his thrilling story."[55] The network agreed in May 1950 to pay Laurence $350 for each of a minimum of four broadcasts. In return, Friendly and other NBC staffers met regularly with Laurence in the latter's New York apartment between late May and the middle of June. Friendly developed the script from those conversations and from Laurence's writings.[56] He said that the goal was not to celebrate the bomb but rather to provide a rejoinder "to the jerk who stands in the streets and yells 'We oughta drop a couple of bombs on them Commies, and quick.'"[57]

Laurence would serve not only as technical advisor but also as the program's conarrator, explaining how the atomic bomb was first developed and what was on the horizon in terms of nuclear weapons and energy. For the primary narrator, NBC and Friendly set their sights high. Bob Hope had hosted a hugely popular radio show for the network for years (much to the chagrin of Norman Corwin and others who had been scheduled against him). His persona was such that he could make the daunting subject of atomic energy seem comprehensible to the average listener. According to Friendly, "We figured people would think that if he could understand it, anyone could."[58] In early June, Hope was sent a confidential letter that Friendly appears to have drafted himself to be delivered under William Brooks's signature. The letter called the new documentary "the finest thing that radio has done since the war" and "the most compelling radio script I have ever seen," having been "written by Fred Friendly, who does the Murrow 'I Can Hear It Now' album." The letter also declared that the script was written specifically with Hope in mind. Little would be required of him: "The entire job is being assembled on tape, and you can do your lines in less than a morning's time, either here [in New York] or on the [West] [C]oast." The letter added a postscript addressing the program's rationale:

> Up to now no one has done anything but try to scare the pants off [the people] about the bombs. As you know, no good decision in the history of civilization has ever been made under the compulsion of fear. It is like a man being frozen rigid by the fear of a snake. He is paralyzed and struck down. We can, through you and the script, show the American people that atomic energy can be as exciting as the way Red Smith [then of the *New York Herald Tribune*] writes baseball. We shall have done a great service to this nation.[59]

Hope recorded his lines for the initial segments of the documentary at the end of June in Hollywood in a single seventeen-hour session supervised by Friendly. According to an NBC internal memo, engineers ensured that their studio setup in California exactly matched that of New York so that each produced precisely the same audio quality: "This will simulate, when edited, a two-way talk between Hope and other characters as though it were done in one studio." In fact, as Friendly boasted to *Time*, "No two people on the show ever saw each other."[60] That included Hope and Laurence, even though the program would be written as a conversation between the two men. Other real-life figures in the program included members of the Enola Gay crew in addition to General Leslie Groves (military leader of the Manhattan Project that created the bomb), Kenneth Nichols (deputy to Groves on the project), and Robert Bacher (a nuclear physicist who also worked on the project). Friendly recorded many people in New York, including an unnamed army official lured to town by an offer of free theater tickets to *South Pacific*. Friendly also traveled to Chicago to record a physicist and to Florida to record the Enola Gay pilot Paul Tibbets.[61]

NBC allowed military officials to check the script for factual accuracy. The Enola Gay crew member William "Deak" Parsons wrote four pages of comments to Friendly after reading a draft. Parsons had devoted himself after the war to countering what he saw as antinuclear hysteria in the media. He told Friendly that in this case he was more concerned with setting the historical record straight, given that certain "war horses" were "very much inclined to allow mythology to be substituted for history as they know it [if] the mythology might be more complimentary to them." Parsons stressed that contrary to what the script draft implied, it had not been a last-minute decision for him to wait until after takeoff to arm the bomb en route to Hiroshima. He made minor revisions to his own dialogue for the production.[62]

A more serious disagreement occurred with Bernard Baruch, the U.S. diplomat whose 1946 speech to the United Nations Atomic Energy Commission had inspired the program's title. "We are here to make a choice between the quick [i.e., living] and the dead," he had declared, before outlining what became known as the Baruch Plan for U.N. control of atomic weapons.[63] Friendly's original draft had Hope approaching Baruch on the latter's usual "bench in Central Park" and asking him to explain atomic energy to him. Baruch was to reply that he could not teach physics and was "not as young as I used to be"; therefore, Hope should consult Laurence instead.[64] However, Baruch apparently wanted to review the entire documentary before he would consent to have his voice used in it. Friendly wanted instead to have an actor impersonate the diplomat, but Baruch vetoed that in communications with Friendly and the NBC chief David Sarnoff. NBC in turn "regretfully" informed Baruch that it would not include him at all: "We certainly wanted the prestige and stature which your participation would have given the programs, but we understand your position and will certainly abide by your wishes." In the end, a public-domain recording of Baruch's 1946 speech was located, and a brief portion was used in the broadcast. A similar recording was used of David Lilienthal, chair of the U.S. Atomic Energy Commission.[65]

Friendly did successfully recruit the actors Paul Lukas and Helen Hayes to play the physicists Albert Einstein and Lise Meitner; the actors received no compensation other than new RCA televisions.[66] Father Wilhelm Kleinsorge had been at a mission in Hiroshima when the city was bombed, and he had become a central figure of John Hersey's book on the attack and its aftermath. A Fordham University priest who had worked at the same Hiroshima mission impersonated Kleinsorge in the part of the program recalling the bombing. To re-create the sound of the first atomic-bomb test in July 1945, Friendly recorded a six-by-eight-foot leather drum being pounded by mallets and whips alongside sixteen turntables all playing thunder at once. He then multiplied the recording eightfold to simulate the explosion's reverberations. Friendly and the engineer Bill Schwartau edited the final documentary with "as many as six or seven tape recorders humming simultaneously."[67]

The Quick and the Dead was scheduled to air in four half-hour segments on consecutive Thursday nights starting July 6, 1950. Press accounts indicated that William Brooks had difficulty at first convincing other network executives to air the series in prime time. Without elaborating, *Time* said that some executives "protested that it was too controversial." *Variety* reported that Brooks and the NBC president Joseph McConnell also had trouble persuading some affiliates to preempt commercially lucrative programming for a nonsponsored public-service series. Finally, "all 167 of [the affiliates] came through."[68] Possibly world events influenced the stations' thinking; the Korean War began less than two weeks before the documentary's debut. One station executive wrote Brooks that the war would "undoubtedly make the public far more aware of the need of knowing more about the Atom and H Bombs. Unless another World War has started before your series begins, I am sure a lot of people will show an interest in the program."[69]

Bob Hope introduced the opening segment: "You're probably leaning back and saying to yourself, 'What on Earth is Hope doing talking about atomic energy?'"[70] He explained that he had realized he "was a very big taxpayer," and "as long as I'm financing [atomic projects], I ought to kind of check up on things." Deciding he needed "a guy that talks atom" the way that "Red Smith writes baseball," Hope went to "the boys in the [NBC] newsroom" (as opposed to Bernard Baruch, as had been originally scripted). In turn, Robert Trout directed him toward Laurence. "He certainly didn't look like a hot reporter or my idea of a Pulitzer Prize winner," said Hope of Laurence. "He was medium height, about fifty-five or so, with a rather thoughtful face and a slight accent." He also had a remarkable story to tell. "It wasn't really my story—it was *man's* story," said Laurence brightly. "I was privileged to be tapped on the shoulder by the army and given the assignment that every newspaperman dreams of."

That assignment was the birth of the atomic bomb. Laurence told of the first test in the New Mexico desert. As a voice reverberating through a loudspeaker counted down the final seconds to the test, Laurence spoke of the bomb as "a two-billion-dollar baby waiting to be born." Then "from the bowels of the earth [came] a light not of this world, the light of many suns in one," followed by a vast mushroom cloud. Finally, "out of the great silence came a mighty thunder," as Friendly's meticulously staged explosion rumbled on the sound track. "How did it happen that this secret had been given to us and not to our enemies?" Laurence asked.

The rumble of the explosion gave way to the ranting of Adolf Hitler. Laurence noted that scientists such as Enrico Fermi were able to escape the fascist powers and emigrate to America. Fermi (impersonated by an actor) told of how he had won a Nobel Prize in 1938 for neutron research. At that point, Hope interrupted to ask Laurence what a neutron was. "Well, here we have to get into physics for a second, Bob," said Laurence. "Let's make it a rule: I'll never give you more than sixty seconds of physics at a time—how's that?" He then explained that the neutron was part of an atom's nucleus and that it could be used to split other atoms. The lesson thus concluded, Laurence continued by describing how the "non-Aryan" Lise Meitner had fled Germany. Meitner (as played by Hayes) discussed her groundbreaking experiments in splitting uranium atoms, and then Einstein (as played by Lukas) read from his famous 1939 letter to President Roosevelt: an atomic bomb was possible, and Germany might be developing one.

The climax of the documentary's debut segment was the first nuclear chain reaction in 1942 at the stadium "where the University of Chicago used to lose all its football games," as Hope put it in one of the occasional wisecracks the script allowed him.[71] Suspense was built through the gradually increasing tempo of neutron counters, indicating that the chain reaction had been achieved.[72] "Some of you may wonder what the purpose of these four broadcasts is," said Hope in conclusion. He quoted David Lilienthal as saying that if "schemers or fools or rascals ever get [atomic energy] out of the people's hands, it may then be too late to find out what it's all about." Hope told listeners that *he* was willing to find out about atomic energy: "How about you?"

Segment 2 of *The Quick and the Dead* covered the rest of the war—the laborious production of uranium-235 at Oak Ridge, Tennessee, and plutonium at the Hanford Site in Washington State; the construction of the bomb at Los Alamos, New Mexico; and finally the Enola Gay mission to Hiroshima. Pilot Paul Tibbets, copilot Bob Lewis, and bombardier Tom Ferebee re-created the flight, with Deak Parsons reading the lines he himself had revised about arming the bomb after takeoff. Then Parsons matter-of-factly described the result: "The great fireball expanded into a huge mass of churning flames. . . . At the base of the lower part of the mushroom was a dust cloud several hundred feet thick which covered most of Hiroshima." Survivors (impersonated by actors) recalled the impact, with Father Kleinsorge comparing it to "a large meteor colliding with the earth." "Bill, I'm staggered and bewildered by the whole thing," Hope said to Laurence. "This was all caused by just a few pounds of uranium being banged together?" Laurence replied that actually less than one gram of uranium had been split, which still had triggered enough blast, fire, and radiation to kill seventy-eight thousand people.

The documentary's third segment focused on the hydrogen bomb. Laurence and Robert Trout recalled three momentous events that had occurred in recent months: the Soviets had detonated their own atomic bomb, the Los Alamos physicist Klaus Fuchs had been arrested for leaking atomic secrets to the Soviets, and President Truman had ordered development of the hydrogen bomb, which Laurence said reflected "the vital necessity to maintain our atomic superiority." Laurence then walked Hope through the physics behind the new weapon—the difference between fission and fusion, the role of deuterium or "heavy hydrogen," and so forth. At the same time, Laurence made clear the weapon's lethalness: "The [radioactivity] would destroy life as it drifted across the Earth's surface. Like Carthage, nothing would ever grow there again. . . . You could destroy every division that fought in Europe with one to three such bombs." Laurence also said that certain unnamed scientists believed that making hydrogen bombs "could bankrupt the nation" due to the high cost of the tritium needed for their manufacture (no exact dollar figure was given).

The segment concluded with the perspectives of two Nobel Prize–winning physicists. First came the actual voice of Harold Urey, whose research had helped make the hydrogen bomb possible and whom Laurence described as "one of the most articulate" spokespeople on the issue. Urey said that the United States had no choice but to try to produce such a weapon, lest its enemies claim that they had built it themselves: "How can we know [whether they are bluffing] unless we have proved to our complete satisfaction that a hydrogen bomb can be built or cannot be built?" Then Paul Lukas was again heard impersonating Einstein and saying that with development of a hydrogen bomb, "radioactive poisoning of the atmosphere and hence annihilation of any life on earth has been brought within the range of technical possibility."

The fourth and final installment of *The Quick and the Dead* was originally to have compared the Baruch Plan with the Soviets' Gromyko Plan regarding nuclear-arms control.[73] Instead, Friendly focused on the positive impact of atomic energy. The segment began with Hope speaking of "the most important date of the atomic age": the day in 1946 that Oak Ridge began shipping radioactive material across the country in the service of medicine. Hope asked Laurence whether "something other than bombs" could come from atomic research. Certainly, said Laurence: "If we can learn to live in the atomic age without allowing this great energy to blow up the world, there is almost nothing the atom can't do for us." The Oak Ridge shipment was proof, as documented by the story of a fourteen-year-old named Johnny who had thyroid cancer. Using the actual voices of doctors, nurses, nuclear technicians, and Johnny himself, the documentary told of how radioactive iodine created at Oak Ridge had helped isolate and treat the boy's cancer at a New York hospital. Similar real-life stories were heard from hospitals and research centers in Boston, St. Louis, Detroit, and Philadelphia, again via the voices of actual doctors and scientists.

"Then Bill, the very rays that killed at Hiroshima may someday save lives?" Hope asked Laurence. "It's already taking place," Laurence replied. Atomic energy, he said, "can turn deserts into blooming gardens. It can air-condition the jungles and make the Arctic waste livable . . . if only we can keep the world from blowing itself up." The documentary concluded with a montage of voices over the steadily increasing ticking of a neutron counter, with Einstein again warning of the possible annihilation of all life, David Lilienthal declaring that God had not given humankind the capacity to learn nature's secrets just to use that knowledge to destroy itself, and Franklin Roosevelt looking forward to a world without fear. Finally, there was Bernard Baruch: "We are here to make a choice between the quick and the dead. That is our business." The clicking of the neutron counter then rose to a crescendo before abruptly cutting off into silence.

NBC promoted *The Quick and the Dead* via a full-page ad in *Variety* on July 12, 1950, and a mailer featuring a picture of a mushroom cloud that was apparently designed to be sent to the press and network affiliates.[74] The network also assembled "atom kits" that, according to *Variety,* included "a piece of uranium ore, a vial of uranium chemical, a sphinthariscope with which you can watch radio[active] activity—and a copy of the script." NBC sent the kits to sixty-five radio editors and critics across the country, although it decided not to send them to major critics such as Jack Gould of the *New York Times*.[75] Regardless, Gould pronounced the program "exciting, informative, and distinguished." Others were similarly effusive. *Variety* said that the documentary was filled with "suspense that might have done justice to a superior thriller," whereas *Billboard* said that it demonstrated what radio could accomplish "as a contributor to America's education and solidarity as a nation." *The Nation* praised it for abandoning the notion "that the average audience age-level is somewhere between ten and twelve." William Brooks enthused that "no single series of this kind has ever produced the results for NBC that this one has," adding, "The only unfavorable review I have seen was in *The Daily Worker*."[76]

Indeed, the Communist party newspaper panned *The Quick and the Dead* as being "aimed at convincing people that the atomic bomb is something to keep." As for the program's pronouncements about needing to keep atomic energy in the people's hands, the paper tartly observed, "The atomic bomb is in the people's hands just as much as General Electric."[77] That was very much a minority view, though. *The Quick and the Dead* went on to win a 1950 Peabody Award for "simplifying and dramatizing a difficult technical subject and [for] dwelling on the good as well as the evil that lies in the conquest of nuclear energy."[78]

The documentary also sealed Friendly's departure from NBC toward the end of 1950. There had been tensions between him and the network the previous year in connection with *Who Said That?* over not only technical problems with the show but also over its scheduling. In addition, the rigors of producing *The Quick and the Dead* had proven grueling for Friendly, causing him to lose twenty pounds.[79] Still, NBC's Brooks had been delighted with the atomic series, and Friendly jocularly responded in kind by saying that he would thank network executives "if I were the kind of man who believed in saying thank you to my employers."[80]

Evidence suggests that Friendly left NBC simply because CBS made him a more attractive offer, including a permanent partnership with Edward R. Murrow. CBS had sent out feelers as early as February 1950 about potentially broadcasting the radio version of *Who Said That?*[81] The following summer, while NBC was airing *The Quick and the Dead*, CBS approached Friendly again. The public-affairs director Sig Mickelson recalled that he and other network executives had grown dissatisfied with the dramatized programs that had been the Documentary Unit's specialty: "It was evident that CBS management felt it had squeezed the radio documentary for all the value it could get and was not interested in scheduling additional programs simply to keep the unit busy." Looking for a fresh approach to such programs by using taped actualities, Mickelson received permission from William Paley and Frank Stanton to recruit Friendly, who agreed that fall to switch networks.[82]

Friendly's *The Quick and the Dead* had drawn significantly upon the older dramatized style of radio documentary. That was evidenced by the scripted dialogue, the presence of Hope and other performers, and the wholesale invention of certain details (such as Hope allegedly seeking out Laurence in his apartment, whereas the two had never actually met). However, Friendly's extensive use of tape pointed toward the future of documentary while building upon what series such as *Voices and Events* already were doing. In particular, the last segment on atomic research and medicine drew upon actuality recordings, many of which sounded less scripted and rehearsed than the segments featuring Hope and Laurence. Hiring Friendly allowed CBS to respond competitively to NBC's new emphasis on recording technology while also tapping into an increased interest in radio news following the Korean War's outbreak.[83]

The move also cemented the Friendly-Murrow collaboration that already had proven so lucrative for CBS's record division. CBS trumpeted the pairing by saying that the two men would employ the same "revolutionary technique" they had used in *I Can Hear It Now* "in a series of projected documentary productions which Murrow and Friendly will do on CBS radio and TV."[84] The reference to television was significant. In his initial meeting with Sig Mickelson, Friendly expressed an interest in eventually producing a "*Life* magazine of the air"—that is, a program centered around pictures. Friendly's new CBS contract allowed him not only to produce radio documentaries but also "to study and engage in certain experiments respecting the format and presentation of television documentaries."[85] The trick now would be gaining the full commitment of Murrow, who had been heard to exclaim, "I wish goddamned television had never been invented."[86]

* * *

An overview of NBC's documentary efforts as the 1940s gave way to the 1950s neatly encapsulates the changes that had occurred since the end of the war. Whereas in 1945, Lou Hazam had written in *Home Is What You Make It* of Americans as "*world* citizens—citizens sprung from a common origin, scattered by fate over a common universal home, and possessed of a common destiny,"[87] his scripts for NBC's *Living* four years later warned Americans of the need to be eternally vigilant regarding the communist threat overseas.

Whereas NBC had clung tenaciously to its ban on airing recordings into 1949, within a year the network president Joseph McConnell boasted that NBC had "instituted the most complete and thorough news department tape recording setup of any network." And whereas radio had been undisputed king at war's end, by the end of the decade McConnell was declaring that "television is taking the country by storm" and that radio would have to "develop still more ingenious and captivating fare for its listeners as television continues to close the gap between its technical perfection and artistic excellence."[88]

Just as CBS's pioneering documentary programs had helped prod NBC into creating documentaries of its own, NBC's innovations in 1949 and 1950 prompted CBS to respond in kind. In addition to luring Fred Friendly away from its rival, CBS also would recruit Elmore McKee and *The People Act* to the network in 1951. The selling points were that CBS by then had overtaken NBC in tape-recording technology and that CBS "offered for discussion several first-class time spots" to air the series.[89] *The People Act* would have its second broadcast run on the new network in 1952. This time it was produced with the support of the Ford Foundation's new Television-Radio Workshop directed by Robert Saudek, who had left ABC to take the job.[90]

The People Act and *The Quick and the Dead* reflected what the historians Kevin Barnhurst and John Nerone have described as the demand for a "firm consensus" on national values during the cold war, reinforcing "the precise meanings of American institutions and their unique superiority to other national traditions, the precise genius of the American character and its unique invulnerability to fascism, the precise classlessness of the American economy and its unique ability to provide luxury for ordinary folk." To be sure, Elmore McKee was hardly the typical cold warrior. Far from shunning the pacifism that he had embraced before World War II, he was still lamenting as of the 1950s that "the family of nations has not yet made a serious effort to earn the organic unity it sorely needs."[91] In that, his sentiments were not far removed from those of Norman Corwin in programs such as *Citizen of the World*. McKee also would later praise Edward R. Murrow's critical 1954 TV report on Senator Joseph McCarthy, saying that it represented a triumph of courage over fear.[92]

Nevertheless, in conceiving *The People Act*, McKee took up the call of liberal anticommunists for a counteroffensive against totalitarians on both the right and the left. The series told of Americans returning to their pioneer roots, rolling up their sleeves, and bettering their lot. Rather than advocating broad-based structural reforms to national problems—reforms that perhaps smacked too much of top-down elitism or, worse, of collectivism—*The People Act* extolled grassroots, bottom-up action typically starting with a small group of concerned individuals. The sociologist Herbert J. Gans has identified such celebrations of "altruistic democracy" as an enduring value of American journalism: "Ideally [in such stories], citizens should help themselves without having to resort to government aid. . . . [T]he news seems to imply that the democratic ideal against which it measures reality is that of the rural town meeting—or rather, of a romanticized version of it."[93]

In a similar vein, *The Quick and the Dead* avoided the worst excesses of anticommunist hysteria; at the very least, it did not assert that "America must have as many nuclear weapons as possible," which Paul Boyer suggested had become the majority media sentiment by 1950. Two years previously, Friendly and Murrow's *I Can Hear It Now* had concluded on a distinctly ambivalent note, with Murrow asking whether the future would bring peace or annihilation. That ambivalence carried over into *The Quick and the Dead*. At a time when opinion polls expressed strong support for building the hydrogen bomb and some periodicals such as *U.S. News and World Report* implied that using atomic weapons in Korea could be a legitimate option,[94] Friendly's documentary reminded listeners of the human toll of atomic warfare, as in juxtaposing the recollections of the Enola Gay crew with those of Hiroshima survivors. The series did not mince words about the destructiveness and financial cost of the hydrogen bomb, and it ended on an ominous note (more so than *I Can Hear It Now* had), with the rapidly increasing clattering of the neutron counter accompanying Baruch's words about choosing between life and death.

That said, the documentary stopped well short of voicing the kind of antinuclear sentiments that other radio programs had done shortly after Hiroshima and Nagasaki. The script had been reviewed by military officials, and there is no indication that they found anything seriously objectionable; it would have been surprising if they had, given Laurence's central role. Far from conceding that he may have compromised himself by being so closely affiliated with the government, Laurence said that he had been "privileged" to be "given the assignment that every newspaperman dreams of." Rather than question building the hydrogen bomb, Laurence stressed the necessity "to maintain our atomic superiority." The following year, in 1951, he wrote that the Korean War had "revealed the extreme danger lurking in any plan to outlaw production and use of atomic weapons in a world constantly threatened by a savage dictatorship," that being the Soviet Union.[95] Laurence's viewpoint had been underscored in the program by scientists such as Harold Urey. In that regard, the *Daily Worker* had been correct—the documentary did imply that the bomb was something to keep.

By including Urey and others, *The Quick and the Dead* to a significant degree also represented a "celebration of the science, engineering, and industry behind the bomb," as the scholar Robert Karl Manoff has described Laurence's reporting.[96] That was due at least as much to Friendly as it was to Laurence: it was Friendly who sought to allay public fears by showing "that atomic energy can be as exciting as the way Red Smith writes baseball" and who helped recruit Hope to make the subject seem more entertaining and less threatening. In addition, Friendly shifted the concluding segment from a discussion of arms control to the benefits of atomic energy. Such rosy depictions had been actively promoted by government and military officials such as David Lilienthal and General Leslie Groves since at least the time of CBS's *The Sunny Side of the Atom* three years previously,[97] and both Lilienthal and Groves were prominently featured in *The Quick and the Dead*. The effect, just as it had been with the earlier CBS documentary, was to suggest that one could peaceably coexist with nuclear weapons and that atomic research could enhance life as much as menace it.

On balance, Friendly's documentary was not an exercise in either pacifism or bellicosity. Its thesis was simply that Americans had to educate themselves about atomic energy, given that they stood at a crossroads where the possible paths led toward either "the quick" or "the dead." That thesis already had become something of a cliché by 1950;[98] it also obscured the fact that citizens had limited power over what to do about the bomb, again as the *Daily Worker* had suggested. There was an implicitly self-congratulatory tone to such programs: we were free to make such decisions for ourselves, whereas citizens of other countries were not. Still, the appeal to collective responsibility and to making informed choices—similar to Elmore McKee's call for "a social climate in which people are held responsible for the world in which they live" and in which community life prospers rather than atrophies—was not to be scoffed at, particularly as such appeals came during the grim opening stages of the Korean War.[99]

They also were in keeping with Edward R. Murrow's philosophy. Prior to *The Quick and the Dead*, Murrow in 1948 had arranged for David Lilienthal to speak to a group of radio executives and urge them to uphold their public-service obligations by providing better information concerning atomic issues. Murrow then commented on the air about the speech: "What Mr. Lilienthal was really saying to the radio executives was this, '[S]tart using your medium, your skill and imagination to inform people about the kind of world in which they live, for if you don't you won't have any security or profits or future.' . . . [F]ew can argue with Mr. Lilienthal's conviction that fear can do no one any good."[100]

The idea that broadcasters could educate citizens to overcome their fears and make rational decisions for themselves would carry over to Murrow and Friendly's new collaboration at CBS. At the same time, such a tenet would be tested as it seldom had been before.

6

Hear It Now

Robert Lewis Shayon had reason to be optimistic that 1950 would be happier for him than the previous year had been. He was "shocked and perplexed" over his firing by CBS while feeling "a sense of public humiliation."[1] At the start of the new year, however, he was offered a new position in Paris supervising the radio operations of the Economic Cooperative Administration (ECA) associated with the Marshall Plan. Shayon was delighted, only to see the job offer suddenly withdrawn. The newsletter *Counterattack* had questioned the appointment, citing Shayon's participation in "Communist front" organizations such as the Progressive Citizens of America (PCA). Soon afterward, the ECA informed him that his services would not be needed because the necessary clearances could not be obtained.[2]

Shayon protested bitterly in a letter: "If I am disloyal, I know of no other American, bar none, who can lay claim to true loyalty." He had in fact worked with the Progressive Citizens of America "at a time when this organization was the hope of many sincere liberals. The principles I found attractive in the PCA were American principles." Still, he had stopped working with the organization by the time it formed the Progressive party that backed Henry Wallace for president. His entire radio career had "been devoted to working for the highest principles on which this country was founded," Shayon wrote. Therefore it was "a great pity" that he was prohibited from working on behalf of the United States "at a time, when above all times, it needs and should call forth the very best and highest in all its citizens."[3]

It did no good; a friendly journalist with sources inside the ECA wrote Shayon that the organization was "in the midst of a battle for funds" from

Congress and thus was not eager to court controversy: "With McCarthy on the war path, I assume that the government agencies, including ECA, are not pressing matters even in those cases where they might like to."[4] "McCarthy" of course referred to Senator Joseph McCarthy, who was charging that the State Department harbored fifty-seven known communists—charges that he first made at almost the exact time that Shayon received his ECA job offer. Shayon's family urged him to go to *Counterattack*'s publishers to try to persuade them to retract the allegations against him, but he refused: "In my view the whole gang of witch-hunters was beneath contempt." A few months later, he was listed in *Red Channels*, and as he later put, "I withdrew from the world."[5]

Red Channels and Blacklisting

The intensifying cold war had emboldened the House Committee on Un-American Activities and anticommunist groups in their investigations of radio. HUAC had subpoenaed Corwin's *One World Flight* scripts in 1947 and held its hearings on the film industry that fall, with members of the so-called Hollywood Ten landing in prison after they refused to answer whether they were or ever had been communists. The rifts that had begun to emerge among liberals by then were demonstrated by Arthur Schlesinger Jr. He scorned what he called "fellow-traveling, ex-proletarian" writers turned "film hacks," similar to his low regard for Corwin.[6]

That said, Schlesinger also had low regard for HUAC, saying that it showed "the dangers to civil freedom of a promiscuous and unprincipled attack on radicalism" through its "reckless accusations and appalling procedures" as well as "the unlovely progeny it has spawned."[7] In effect, *Counterattack* was one of those progeny in that its publishers served as their own ad hoc investigating committee. Three onetime FBI employees formed American Business Consultants in the spring of 1947 and soon began publishing a weekly newsletter purportedly exposing communists in American business, including broadcasting. In December 1947, they attacked Corwin for "tender, unswerving devotion to the Kremlin." By 1949, *Variety* reported that fears of being labeled procommunist had "grabbed a stranglehold" on broadcasting to the extent that "any actor, writer, or producer who has been even remotely identified with leftist tendencies is shunned." According to the CBS executive Sig Mickelson, *Counterattack* "was on the desk of virtually every advertising agency executive and radio and television network official in the New York City area." If Shayon's experiences with the aborted ECA appointment were any indication, the newsletter's influence may have extended to the nation's capital and even overseas.[8]

CBS was a particular target of *Counterattack*. The newsletter declared in July 1949 that "'NBC and Mutual are least satisfactory to Communists, that American Broadcasting Co. is about at halfway between most satisfactory and least satisfactory, and that CBS is tops as far as Communists are concerned.'"[9] Not only had the network long been viewed as a haven for liberals (Corwin and Shayon among them), but it also had sought to differentiate itself from NBC after the war by developing and owning its own programs and putting on its payroll a significant stable of writers and actors, some of whom inevitably fell under suspicion. Soon after *Counterattack*'s broadside against CBS, the network attorney Joseph Ream contacted the newsletter's copublisher, Theodore Kirkpatrick, to discuss how to begin ridding the network of purported subversives. Corwin and Shayon were already gone by then, but by the start of 1950, both men's names appeared on a lengthy list of "Most Undesirables" that was circulated among top CBS executives—none of those cited was to be used even as a freelancer due to "their record of affiliation with communist fronts and causes."[10]

Red Channels was *Counterattack*'s magnum opus and effectively codified a Most Undesirables list for the entire broadcast industry. Published in June 1950, the 215-page book included an introduction by Vincent Hartnett, who had joined forces with the newsletter's original publishers. Hartnett wrote that *Red Channels* sought to counter those who were spewing "pro-Soviet, pro-Communist, anti-American, anti-democratic propaganda" via radio and television to an unsuspecting public. "A few documentary programs produced by one network in particular have faithfully followed the [Communist] Party line," the introduction declared.[11] That network went unnamed, but any doubts as to its identity could be erased by perusing the annotated list of 151 names that followed. They included Corwin, Shayon, Robert Heller, William Robson (who had written *Open Letter on Race Hatred* during the war and accompanied Shayon to Europe), and Arnold Perl (the author of *The Empty Noose* and *Fear Begins at Forty*); all had worked for CBS. Among the others listed were Luther Adler, the star of Shayon's *The Eagle's Brood*; Lee J. Cobb, the narrator of Corwin's *Citizen of the World*; Clifford Durr, the Blue Book supporter and former FCC commissioner; Tom Glazer, a collaborator with Erik Barnouw on the antisyphilis project; Ben Grauer, the narrator of NBC's *Living* series; Will Geer, the former *Living* actor; Joseph Julian, the star (ironically enough) of *Communism—U.S. Brand*; and Norman Rosten, the letter-writing nemesis of Morton Wishengrad.[12]

Red Channels' authors fervently believed that their cause was just. Theodore Kirkpatrick appeared on a radio talk show to defend the book a couple of months after it was published. "He carried himself quietly, soft-voiced and completely assured for the most part," said one observer at the time, "as religiously righteous as any zealot sure of his moral stand, with mind apparently closed to any counter opinions or arguments."[13] Still, *Red Channels* claimed not to wish to foster hysteria: "In screening personnel every safeguard must be used to protect innocents and genuine liberals from being unjustly labeled." Some of the language in the book's introduction in fact recalled that of anticommunist liberals like Schlesinger and Wishengrad. Issues such as "'academic freedom,' 'civil rights,' 'peace,' the H-bomb, etc." were "perfectly legitimate in themselves." The concern was when "well-intentioned 'liberals' [were] exploited" via their participation in such causes to "point up current Communist goals."[14]

The publishers of *Red Channels* could not by themselves keep those whom they deemed dangerous off the air. That was exemplified by Ben Grauer, the sunny-voiced announcer for several NBC programs besides *Living*. Grauer had been identified with a number of liberal causes and in 1946 had helped lead a coalition within the American Federation of Radio Artists to defeat an anticommunist resolution that the coalition saw as "a union-splitting tactic."[15] At the same time, he was a highly visible presence on NBC. In a brief 1947 piece headlined "Grauer vs. Grauer," *Variety* needled him for one day telling a "'crisis-in-radio conference'" that there was "very little freedom left on the air" (a viewpoint consistent with that era's media-reform movement) and then the very next day informing an NBC Symphony radio audience that the existing system presented "truly free radio" devoted to "pleasing its great audience" as opposed to the dictates of "arbitrary critical groups." Whatever embarrassment that might have caused Grauer, it suggested that he was savvy or malleable enough to be able to protect his career. *Red Channels* seemed to have no serious effect on him; NBC even sponsored a "Ben Grauer Day" in New York not long after the book appeared in 1950. Coincidentally or not, that event occurred at about the same time that Grauer was lending his voice to *Living*'s "Malice in Wonderland" series, which urged proactive American resistance against Soviet propaganda.[16]

Another *Red Channels* listee, Gypsy Rose Lee, similarly escaped lasting damage thanks in part to the stand taken by Robert Saudek and other ABC officials. The Illinois American Legion protested Lee's hosting of a new radio quiz show on the network, calling her a "dear and close associate of the traitors of our country." The network asked for proof of the allegations, and when none was presented, it allowed Lee to host the show as scheduled. As a result, ABC was presented a special Peabody Award in 1951 for having "refused to be stampeded" into "either firing or refusing to hire writers and actors on the basis of the unsupported innuendoes contained in a publication known as 'Red Channels.'"[17]

Such displays of courage or common sense were comparatively rare, however. *Red Channels* could have an insidious effect, even on those who were wholly sympathetic to those listed. Erik Barnouw had worked with many of them besides Glazer; he also had edited the 1945 anthology *Radio Drama in Action*, which had featured the work of no fewer than nine listees.[18] It was not surprising that he later described the names in *Red Channels* as "a roll of honor." Nonetheless, he recalled in his memoirs that when he first saw the book, he experienced an odd blend of relief and guilt over not having been listed himself, along with a disquieting suspicion: "Did I *really* know [the people listed]? Were there things about them I did *not* know that had landed them in this strange volume? To find myself asking such questions, even to myself, was chilling. The poison was spreading."[19]

Further evidence of the toxin's spread was that *Red Channels* quickly disappeared from New York City bookstores, the copies apparently snapped up (as *Counterattack* had been) by broadcasting and ad-agency executives. The book appeared at a frightening time; apart from the Soviets developing atomic weapons, the communists controlling China, and the Alger Hiss espionage case making headlines, the Korean War erupted virtually simultaneously with the book's publication.[20] For all that, as one historian has since written, the "clammy fear" among the media executives was "not of totalitarianism but of the prospect of lost short-term profits."[21] That was especially the case after a grocer, Laurence Johnson, and his family launched a campaign urging companies not to sponsor programs that featured allegedly tainted performers while also boycotting the products of those companies that did sponsor such programs.

CBS was among the most fearful of all. At the end of 1950, it imposed a loyalty oath, asking all its employees whether they had ever been members of the Communist party. Many network staffers protested, and Jack Gould in the *New York Times* pointed out how preposterous it was to expect subversion-minded communists to expose themselves by affirmatively answering such a question. Gould also said that the oath raised "the disquieting specter of one citizen assuming the authority to investigate and pass judgment of another."[22] Again, though, money was the real concern. As the administrator of the CBS loyalty oath, Daniel O'Shea, later recalled, "It wasn't that I felt myself in the middle of ridding the world of Communists—rather of some group or other who were affecting CBS's business."[23]

That cast into the wilderness a substantial number of those cited in *Red*

Channels. Some in fact had for a time adhered to the Communist party line to the point of defending Stalin's purges and the Hitler-Stalin Nonaggression Pact, a stance that, however wrongheaded, had not been illegal.[24] Many others never had held communist sympathies whatsoever. Norman Corwin felt compelled to defend himself in a written statement. "If I were a communist or fellow traveler, I had one of the best platforms in the world," he wrote:

> I had the ear of millions of Americans—sixty-three million in one night on one occasion. I had no interference in the form of censorship and none of the usual checks of supervision. I could say what I wanted . . . [and] I would defy anyone to find anything advocating a foreign system of any kind superior to ours. I defy anyone to show that in whatever allusions I ever made to the forms and ideals of our government, as apart from the occasional abuses of those ideals, I had done anything but strengthen our respect and deepen our appreciation for the United States, its laws, its people, its constitution [25]

In response to a query from the executive director of the American Civil Liberties Union, Corwin took issue with the specific activities that *Red Channels* had attributed to him. He had not even attended some of the suspicious events he was supposed to have attended. Corwin also said that the *Red Channels* listing had not affected him as much as it might have had he regularly accepted commercial sponsorship. Regardless, he was still "graylisted," which meant fewer work offers during what he later recalled as being a "miserable time" when he saw colleagues "exiled, punished, jailed, ostracized."[26]

For the actor Joseph Julian, as he later wrote, "two citations in *Red Channels* cost me my livelihood for over three years." One was for participating in a meeting during the war calling for the opening of a second front in Europe (which would have relieved pressure on the then-besieged Russians); the other was for attending a rally headlined by the movie star Burt Lancaster calling for the abolition of HUAC. The only regular work Julian could find was in industrial and training films. Bizarrely, those included one made solely for the eyes of a congressional committee regarding how the military would function during an enemy attack, with Julian donning a general's uniform and filming inside the underground headquarters of the Strategic Air Command. Meanwhile, he harbored secret anxieties about why no one else would hire him. "If the odds were ninety-nine to one that it was because of *Red Channels,* actors would dwell on the one percent possibility that it was their own fault," he recalled. "It led to a terrible erosion of faith in ourselves." Julian fantasized about hunting down actors'-union officials who supported blacklisting and hurling them under a subway train: "I was shocked to find myself speculating on how many broken lives it might redeem."[27]

Morton Wishengrad, whose *Communism—U.S. Brand* had featured Julian in the lead role only two years earlier, was appalled. In a letter to the editor of the liberal anticommunist *New Leader* magazine, he decried Tom Glazer and Jean Muir's listings in *Red Channels*. (Muir, who had been abruptly dismissed from the NBC series *The Aldrich Family*, was one of the most notorious blacklisting cases.) Wishengrad also attacked the premise behind the book:

> When the *New York Times* reports that the publisher of *Counterattack* "indicated that one method for a person to absolve himself of Communist charges was to testify publicly on how he might have been duped, if that were the case," I see the Stalinism of purge, confession, and public penance demanded not by the Government of the United States, but by a private government, self-proclaimed, self-righteous. . . . Yes, of course, a majority of those listed in *Red Channels* are undeniably Party members or fellow-travellers, but I hate to see such a listing. I want to fight Communism with the methods approved by democracy, not by methods which inevitably result in the forced collectivization of ideas and the compulsion or even extortion of avowals.[28]

Wishengrad similarly defended the actor Sam Jaffe against his *Red Channels* citation, and when Joseph Julian sued *Counterattack* for libel, Wishengrad (along with Robert Saudek and Edward R. Murrow) served as a character witness. When the suit came to trial in 1954, though, the judge dismissed the case before it could reach the jury. That ironically worked out well for Julian in that it seemed to persuade media executives that if *Red Channels* had not actually branded Julian as a communist (which is what the trial's outcome implied), he in fact really must not have been a communist. Soon he began receiving work offers again.[29]

Robert Heller's fortunes took a different turn. At first his listing in the book seemed to have no bearing on his career. He retained his executive's post at CBS and even was mentioned as a potential candidate for promotion to network vice president. Then came the imposition of the CBS loyalty oath in December 1950. The public-affairs director Sig Mickelson recalled that between Christmas and New Year's Eve, he found a note on his desk from Heller apologizing for leaving without saying goodbye. When Mickelson checked Heller's office, he found it empty: "I assumed that his sudden departure was in some way connected with the citations in *Red Channels* or perhaps the CBS 'loyalty oath,' but no one was talking."[30] At the time, it was reported that Heller had taken a position with Louis Cowan's production company. *Variety* reported that it was the result of "many weeks of negotiation which culminated in a contract a couple of weeks ago," and *Billboard* indicated that he would "function as a top creative programming brain, with a royalty cut of every show he puts together." Abruptly, the Cowan job fell through, and Heller instead was said to be working on a scenario for an unnamed film company. That too apparently came to naught. Heller finally left the country altogether, spending three years producing radio and television shows in Mexico before landing a position in British television in 1954 with the help of a recommendation from Edward R. Murrow.[31]

Robert Lewis Shayon continued to resist his family and friends' entreaties to try to clear his name with *Red Channels*' publishers. In October 1950, in the midst of a debate over communist influence in the Radio Writers Guild, he joined Erik Barnouw, Morton Wishengrad, and other guild members in signing a petition declaring opposition to "blacklists and morals clauses whatever their source," as well as to "Communism and all other 'total' doctrines."[32] That did not help Shayon find work in broadcasting. He did take a position as a radio and television critic for the *Saturday Review* magazine, but he sank into a morose funk as his savings dwindled. Finally, in 1953, he swallowed his pride. Taking the advice of peers in the radio industry, he prepared a statement under the heading "To Whom It May Concern" to be circulated to the FBI, American Legion, and other organizations identified as anticommunist.

Shayon's statement addressed each of the three *Red Channels* citations against him. He indeed had signed a 1945 reelection petition on behalf of the New York City council member Benjamin J. Davis, who was African American and a communist. However, Shayon said that he signed it only because he thought that Davis was the victim of racial discrimination. The other two citations echoed the charges that *Counterattack* had made against him at the time of the ECA appointment—Shayon had signed a 1947 letter by the Progressive Citizens of America that denounced Hollywood's capitulation to HUAC, and he also had spoken to a PCA conference that October. In that 1947 talk, he had defended the freedom "of minority opinion to criticize, to challenge, to experiment in its proper quest for new phases of political, economic, and cultural truth" while resisting "the negative, unthinking, hysterical fear that moves the enemy"—that is, red-baiters and blacklisters.[33]

Now, six years later, Shayon sought to clarify his thinking:

> I have always sought to defend our fundamental democratic principles, and to extend their practical applications constitutionally. In so doing, I associated

myself on a number of occasions with groups that attracted other like-minded individuals: but it is now clear to me that those groups were used by Communists in varying degree to serve their own destructive ends. I believe that the balanced record of my actions proves that I have never been a Communist or Communist sympathizer; but that unintentionally and unwittingly, I helped the Communists in their policy of creating and intensify[ing] conflicts and tensions in our society.

Communism, in my opinion, is evil materialism. It masquerades under the dream of a materialistic Utopia, but it carries within itself the seeds of tyranny, cruelty, and the extinction of the very liberties it claims to espouse. The Communist Party in America is clearly an agent of a foreign power.

I am confident that fair-minded Americans will think it unjust for me to [be] deprived any longer of the opportunity to employ constructively whatever gifts I have—by virtue of a cause to which I never subscribed—and by which I will continue to oppose with all the vigor I command.[34]

Shayon took the statement to Vincent Hartnett, who had written the introduction to *Red Channels*. "You never should have been in that book," Hartnett told him.[35] Hartnett followed up with a "supplementary" report that he sent to the grocer Laurence Johnson and the networks. Regarding Shayon's 1947 PCA talk, he wrote, "All available evidence indicates that Mr. Shayon, a liberal, was 'used' by the Communists on this occasion, since the overall purpose of the conference was to agitate for the actual abolition of [HUAC] and to prevent any curbs on Communist penetration of the cultural field." Simultaneously, *Counterattack* crowed over Shayon's capitulation: "*SHAYON has issued a statement* that reveals how 'Red Channels' has been instrumental in bringing him around to thinking straight about Communist issues." Referring to *Red Channels*' introduction, which spoke of well-meaning liberals being turned into communist dupes, the newsletter concluded that the book "helped to clarify SHAYON's thinking about what speaking before, and associating with Communist fronts, actually means."[36]

Shayon went on to a long tenure as a critic and communication professor, but his bitterness over his blacklisting lasted for decades. He was convinced that *Red Channels* had led directly to his ouster from CBS and destroyed his broadcasting career. It was only when he was writing his memoirs late in his life that he realized that the red-baiting book had not appeared until the year after he was fired and that William Paley simply "had made a routine business decision" by cutting costs in radio to help underwrite the transition to television. "I was neither victim nor hero," Shayon concluded. "I was a human being, as all of us were, trapped in the tangle of peculiar times."[37]

Murrow, Friendly, and *Hear It Now*

Edward R. Murrow negotiated the odd times by picking his battles carefully. Shayon wrote in his memoirs that although Murrow did defend some CBS staffers and others against red-baiting, "he never, so far as I know, protested the Communist charges against me. He never spoke to me after my firing."[38] Nor did he fight the CBS loyalty oath, even though some of his colleagues were prepared to do so. When one of them, the reporter Bill Downs, said that he would never sign it, Murrow told him that he had no choice. Sig Mickelson recalled that Murrow's support for the oath helped defuse the protests against it.[39]

Murrow also was careful concerning what he said on the air, distancing himself from whatever implicit advocacy had seeped into his wartime reporting. After stepping down from his administrative post, he began a nightly newscast on CBS in 1947 with the promise "not to use this microphone as a privileged platform from which to advocate action." That is not to say that he was completely neutral. At the time of the HUAC hearings on Hollywood, he told his listeners that "either we believe in the intelligence, good judgment, balance, and native shrewdness of the American people, or we believe that government should investigate, intimidate, and finally legislate." He added that the right to dissent had been the first to vanish "in every nation that stumbled down the trail toward totalitarianism." Similarly, when Joseph McCarthy was first making his charges concerning alleged subversion in the State Department, Murrow approvingly pointed to a *New York Times* editorial tweaking the senator, thanking the newspaper "for reminding us in these rather hysterical days that 'guilt by association' is, in the true sense, an un-American doctrine."[40]

Still, as one of his biographers later observed, Murrow "was a conventional anti-Communist" who belonged to the Committee on the Present Danger, which backed the American foreign policy of containment against communism. A leftist newsletter condemned his broadcasts as "red-baiting rhetoric."[41] When he produced a pilot for a radio documentary series in 1949 titled *Sunday with Murrow*, he included a commentary on "how to control Communism without endangering the right of dissent." Acceptable measures for doing so included "not only the exposure of Party membership lists (coupled with heavy penalties for failure or faking), but also exposure of sources of financial support." That would make it highly unlikely that any communist "would be allowed employment in government or critical industries."[42]

The pilot never aired; according to one account, CBS executives and potential sponsors believed that there were "'too many' news documentaries on the air."[43] By late the following year, though, Fred Friendly was on board with the mandate of revitalizing the network's documentary programming via recorded actualities. His first program with Murrow was a half-hour show in November 1950, *Report to the Nation—The 1950 Elections*. A reviewer praised it as being "a skillful autopsy on the recent election campaign with the tape machine used as a revealing scalpel." The program exposed the electioneering stunts to which candidates had stooped and as such represented "a first-rate anthropological study in primitive thinking."[44]

That set the stage for the new weekly radio series that Murrow and Friendly would coproduce starting the following month. *Variety* reported that the new program originally was intended to be only a half-hour before it "caught the fancy of board chairman William S. Paley, who okayed its expansion to a full hour under the Murrow–Fred Friendly production aegis, with a hefty budget to boot." In a separate story, the show-business weekly went so far as to say that Paley had actually "conceived" the program.[45] Paley often claimed credit not due him, but he did regularly help shape CBS's programming (as he had done with the formation of the original CBS Documentary Unit), and he enjoyed a warm relationship with Murrow, in marked contrast to that between Murrow and other CBS executives such as Frank Stanton and Sig Mickelson.[46]

Significantly, *Variety* also reported that "the web is already thinking in terms of a comparable show for TV." For the moment, though, it would be strictly a radio venture. It was scheduled during what the *New York Herald Tribune* critic John Crosby called "an excellent and highly salable time" on Friday nights, "at no little sacrifice" for CBS. That included substantially rearranging the network's broadcast schedule to accommodate the new program as a sustaining, nonsponsored series. Network publicity said that the series would employ all of CBS's resources and keep "four recorders going day and night" in a tape room "for exhaustive coverage of the news." At first, the series was to keep the title *Report to the Nation,* but just before its debut, it took a new name explicitly linking it to the highly successful record album that Murrow and Friendly had produced—*Hear It Now*.[47]

CBS compared the series to "a weekly news magazine" devoted to "every facet of today's living," recalling Fred Friendly's originally stated desire to produce a "*Life* Magazine of the Air." It would have regular contributors: Don Hollenbeck presenting media criticism along the lines of his former program *CBS Views the Press,* Red Barber covering sports, Abe Burrows reviewing theater, and so forth. The series also was to offer an original musical score each week by an eminent composer. Finally, it would profile a prominent figure in that week's news.[48]

The series debuted on 173 CBS-affiliated stations on December 15, 1950. It began with an audio montage of newsmakers interlaced with a David Diamond musical theme that would be reused on Murrow and Friendly's television series *See It Now*. An announcer stressed that "all the voices and sounds you will hear are real and are presented as they were spoken in the heat and confusion of a world in crisis," with the hope that "the collection of these scraps of sound into a weekly recorded history may add another dimension to our understanding in the difficult days ahead."[49]

Indeed, *Hear It Now* premiered at a particularly difficult and crisis-laden moment. A Chinese counteroffensive in Korea was threatening to force U.S. troops off the peninsula. The program featured sounds of the battle combined with Murrow's narration ("That explosion was *in*coming mail!") and an actuality from a wounded marine: "The Chinese were around us like bees. There was a million of them at least. How I got out . . . I'll never know." The show also included excerpts from a United Nations debate over the war, including China's delegate. Murrow branded him "a representative of the hordes that had poured down from the Yalu River" and scornfully described his "staccato, high-pitched voice [that] blasted out at the United States and the West in a barrage of words and unending paragraphs."

The profile of the week highlighted General Douglas MacArthur, whom Murrow called "one of the most dominant and controversial" figures of his time. CBS had cabled MacArthur and his press chief in an unsuccessful attempt to have the general record a statement for the program.[50] As it developed, the profile of MacArthur relied mostly on archival recordings. Murrow took no clear stand on the general but did point to MacArthur's promises that his military strategy would produce peace, which obviously was far from being realized.

Alongside news of Korea and the domestic reaction to it (including a Montana draft board that went on strike because the United States had not used the atom bomb against the communists), Red Barber reported on the ouster of the baseball commissioner, Abe Burrows critiqued a Broadway revue, Bill Leonard (later to become CBS News president) praised the film *Born Yesterday,* and Don Hollenbeck scolded President Truman for his "astonishingly indiscreet" letter blasting the newspaper critic Paul Hume after Hume panned Truman's daughter's voice recital. The program also featured Carl Sandburg reciting from *The People, Yes.* Murrow said that the poet was asked to read "because this is a time for great oratory or great wisdom, and we seem to have little oratory to brace us these days."

Hear It Now's premiere received a 10.5 Nielsen rating, which Murrow seemed to find mildly disappointing, but which CBS trumpeted as representing ten million listeners.[51] *Billboard* praised the show as "alternately stirring, grave, humorous, and provocative" and a riposte to "the sad sacks who have been holding their own private wake over the still warm body of radio." Others had reservations. Jack Gould in the *New York Times* charged that the debut had been "abominably organized," adding that "when *Hear It Now* learns to relax it should be a vastly improved program." For John Crosby in the *New York Herald Tribune,* the series was "one of the finest ideas to come up in a long time." However, it needed to be more critical and "take off its gloves and swing." Sentiments similar to Crosby's came from Robert Lewis Shayon, who reviewed the program in the *Saturday Review*. "It is a good thing to be using actuality tapes," wrote Shayon, but they should be coupled with an "honest, creative, responsible, and courageous" editorial viewpoint. Implicit in his critique was that *Hear It Now* ought to be more like his own CBS documentary *The Eagle's Brood,* which had not used actualities but had vigorously sought to mobilize listeners.[52]

Murrow mostly took such comments in stride, saying that it was necessary to "make your mistakes and get some informed criticism." On the air, he said that *Hear It Now* was "still experimenting."[53] In a letter to a sympathetic listener, though, he disagreed with arguments that the program should be more outspoken: "It may be that at times we do nothing more than contribute to the confusion of our fellow countrymen, but at least we give them their confusion raw." He echoed Friendly, who said that they sought to use tape creatively and "let people listen to the raw stuff."[54]

In fact, *Hear It Now* in succeeding weeks would grow more adventuresome in its sound experiments, while Hollenbeck, Leonard, Burrows, and Barber all would be quietly dropped. Burrows had been listed in *Red Channels,* which threatened his continued employment at CBS until he signed a statement of contrition similar to what Shayon later would write. Hollenbeck also would become an infamous victim of red-baiting, whereas Leonard was one of the CBS staffers who had vociferously protested the loyalty oath before finally signing it.[55] Nonetheless, there is no evidence to suggest that their departure from *Hear It Now* was politically motivated (certainly Red Barber, who was dropped at the same time, was not a controversial figure). The series also abandoned any aspirations of presenting an original musical score each week;

Diamond's theme would be heard at the start and close as well as during a station break, but all other music in the series would be indigenous to the stories being presented. At the same time, a cautious but distinctive editorial stance would emerge.

Hear It Now's staff included Joe Wershba and Ed Scott (who would both also work on *See It Now*), John Aaron and Jesse Zousmer (who would co-produce Murrow's television show, *Person to Person*), and Irving Gitlin (who would become a top documentarian for CBS and NBC). They worked all week to prepare each Friday program. CBS reporters and network affiliates sent tape to New York via plane or closed circuit; audio also came from the BBC and the Voice of America. Friendly shaped the various stories into a rough cut, while Murrow worked on his narration, tying together the elements. Following a Friday rehearsal and the insertion of any breaking news, the final program was then taped to be played back on the air shortly afterward, with a backup copy ready in case the master tape went haywire.[56] According to one observer, "Murrow made the principal decisions as to contents and order of precedence," while "Friendly was responsible for the close, meticulous editing of the program." Although Murrow restrained Friendly's more extravagant impulses, there still was plenty of room for irony and invention.[57]

A month into *Hear It Now*'s run, Murrow ended the broadcast with a statement of purpose: "Just as we believe that often one picture is worth a thousand words, occasionally one word or one sound is worth a dozen pictures." That week's show provided vivid examples of their philosophy in action, including a piece on the national budget. "How do you translate the budget of the United States into sound?" asked Murrow. The answer was to record the roar of a federal printing press that made one-thousand-dollar bills at the rate of 180,000 per minute. Even at that speed, the press still would have to work eight hours a day for two and a half years to print enough bills to cover the entire budget, said Murrow. Then, to show how the budget was allocated, he took a hundred pennies in hand and gradually dropped them before the microphone. Health and education, the census, and slum clearance received four cents of every dollar (*plink, plink, plink, plink*). In contrast, national defense received fifty-eight cents, which Murrow underscored with a long, clattering fistful of coins.

That same program featured a freewheeling treatise on the common cold, which Murrow said that he was suffering from at that very moment. A montage of radio commercials promising cold relief was followed by another montage of people offering their pet home remedies. Humans and apes are the only two species that come down with colds, said Murrow—and then there was a gorilla sneeze. Natural sound from a vinegar factory was also heard; colds were relatively rare there, which a foreman credited to fumes killing the germs. The final word was given to a doctor at Harvard: Get plenty of rest and fluids.[58]

Hear It Now continued airing features using ample sound and Murrow's spare narration. It followed the boxer Sugar Ray Robinson before a fight with Jake LaMotta (which many years later became the gory centerpiece of the film *Raging Bull*). Robinson was heard Valentine's Day shopping before heading to the arena and pounding LaMotta to a pulp as Robinson's mother cheered him on from ringside: "Hit him *any*where!" A report on baseball's 1951 opening day featured an exchange between Joe DiMaggio and a shyly nervous Mickey Mantle, about to start his rookie season. A Memorial Day feature included everything from "Taps" played at the Tomb of the Unknown Soldier to a father in holiday traffic bickering with a child needing a bathroom: "What am I supposed to do, *make* one?" The same week, the program profiled twenty-four hours at the *New York Times*, with sound from editorial meetings, the composing room, and news dealers hawking papers on the streets.[59]

Hear It Now drew upon CBS's overseas correspondents as well as its own staff. To mark one holiday, it featured a playful if at times patronizing story on women's hats: "Easter just isn't Easter without a new bonnet—at least that's what our wives have been telling us," said Murrow. From Paris, David Schoenbrun reported on a new hat design debuting on the fashion runway. That was followed by a recording made by Ed Scott in a New York store in which a salesperson sold a knockoff of the same design for $10.95. The salesperson assured her customer that she looked "chic" (which she pronounced as "chick"). In another program, Scott and Joe Wershba were sent into the Blue Ridge Mountains to record an "oral document" of "Moonshine USA." Along with actualities from "Antiseptic Sam" and "Rattlesnake Pete," listeners heard a federal raid and the blowing up of a still.[60]

The moonshine segment was offered as an ironic counterpoint to a major domestic story of the time, the Kefauver Committee hearings on organized crime. At the segment's start, a committee member was heard declaiming that the Mob did not represent America and that those wanting to know what the country was like should "go out in the hinterlands" where people "pay their taxes." *Hear It Now* covered the Kefauver hearings, with excerpts of the crime boss Frank Costello parrying Senate queries ("I don't answer no trick questions"), followed by stories of how TV had turned the hearings into the "great new hit show in the land." A cinema owner lamented the decline in business, and a man complained that his wife was too engrossed in the telecasts to do housework.[61]

Just prior to the Kefauver hearings, a point-shaving scandal had rocked college basketball. "This occurred in a climate where consciences have become calloused, where the dollar is the big symbol," said Murrow. He added that America's youth were on the whole strong and would meet their test "if they get the leadership they deserve." In a later report on juvenile delinquency, Murrow condemned those who persisted in "passing the buck," adding, "We are the ultimate buck. If we continue in our failure to enforce our laws, the responsibility for the degradation of our youth will be ours."[62]

Hear It Now also reported on the country's rising divorce rate and touched on the nascent civil-rights struggle, with Ralph Bunche condemning southern segregation. No domestic story received more attention, however, than the home front's response to the Korean War. There was a Christmas segment on how Peoria, Illinois, was reacting to the national emergency and a similar report on Detroit, the "heart of industrial might." In response to war-related inflation, the series featured the "biography of a pound of steak," tracing it from a Montana ranch and an Iowa feed farm through the Chicago stockyards and finally to a New York City butcher, with all parties disavowing blame for higher meat prices.[63]

Murrow was especially keen that citizens assume responsibility regarding the war overseas. "Has any greatness been demanded of you recently?" he asked his listeners. They could not bring their troops home for Christmas, he said, but they still could "as free men and women accept and welcome the demands" being made of them. To mark the fortieth anniversary of the Triangle factory fire, *Hear It Now* reported on the New York City garment industry at a time when labor-management clashes threatened to disrupt wartime mobilization. Expressing sentiments similar to those in *The People Act*'s program on the garment business that aired on NBC at about the same time, Murrow said that "compromise and reason" had dramatically improved working conditions while also demonstrating the importance of good labor relations: "The fabric of this country has been woven by all of us. . . . It's a good garment, not completely finished, needing constant attention, but still the envy of our neighbors. We ought to wear it with pride and assurance."[64]

The biggest story of all was the war itself. CBS's Korea correspondents included George Herman, John Jefferson, and Robert P. Martin; rather than voice stories themselves for *Hear It Now*, they usually sent raw tape to New York to be fashioned into pieces that Murrow narrated. Occasionally, though, they recorded on-scene reports for the series. Robert Pierpoint, then only twenty-five and new to the front, produced what Murrow called a "rare record of a human being's indoctrination to combat," highlighted by a "friendly" artillery round falling short of its target and exploding near where Pierpoint had hurriedly taken cover.[65]

The coverage encompassed the grimmest stages of the war. By January 1951, the Chinese had overrun Seoul. Over sound of the mass flight ahead of the Chinese advance, Murrow described the scene: "The lonely, lost children looking vainly for their parents. The dead rotting on the roadside. The smell and the frozen dust of battle and retreat and disaster everywhere. And everyone heading south." Later, under General Matthew Ridgway, the tide was turned, and Seoul was retaken. By Easter, after a period in which Murrow said that "our faith in military leaders and in ourselves as a people was badly shaken," there was reason for optimism: "In addition to all the deep religious significance of Easter, it marks the beginning of good fighting weather. It is a season suitable not only for hope, but for courage."[66]

Early in the war, Murrow had traveled to Korea to see the war firsthand, and he filed a bleak report asking whether "serious mistakes" had been made and whether the war would only drive Korea further toward communism. CBS had incensed Murrow by not airing his report on the grounds that it could hurt the war effort.[67] On *Hear It Now*, other than the oblique reference to shaken faith in military leaders, Murrow hid any reservations he may still have harbored about the conflict. He made no pretense at impartiality, discussing U.S. forces in "we" and "our" terms while describing the other side as "a fanatical enemy attacking and dying." Some of *Hear It Now*'s references to the Chinese recalled those aimed at the Japanese during the previous war. Murrow summarized one U.S. battlefield communiqué: "'They're coming at us like fleas, [and] we're killing them like fleas.'" Soldiers interviewed after a particularly bloody clash described counting the dead "chinks" piled before each gun emplacement so that proper credit could be given to those manning the guns.[68]

If the enemy was dehumanized, the American serviceman was lionized. *Hear It Now*'s grandiose "profiles of the week" gave way to the "little picture" stories that *See It Now* later made famous.[69] One program took listeners to a "tiny hamlet" in South Carolina to hear friends and family of a private who had been in the thick of the fighting and whose current status was unknown. Another program visited the Memphis parents of a young soldier who had won the Distinguished Service Cross: "He happened to be a Negro," said Murrow. A Medal of Honor recipient from Oklahoma who had been killed in action was similarly profiled. There also was a lengthy piece using what Murrow called the "smallest, most inconspicuous equipment we could get" to record the tenderly awkward reunion of a Korea veteran with his family in New Jersey, including the infant child he had never seen.[70]

The wounded received attention as well. Robert Pierpoint flew with a helicopter crew transporting casualties to a MASH field hospital. Another segment began with the sound of sawing. "No, they're not cutting wood," said Murrow. "They're cutting off a soldier's arm at Walter Reed General Hospital. That's not a very pleasant sound. But then war isn't very pleasant, either." A report followed on how the hospital rehabilitated military casualties of the war.[71]

No *Hear It Now* piece generated more response than its "biography of a pint of blood." The story began with Murrow warning that it would "use sounds and voices franker and less temperate than those usually heard on the radio. We believe that too much is not *ever* said on the radio." That included the "brutal naked sounds and phrases that our sons and brothers hear every day on the battlefield." The story then followed a pint of blood from a donor's arm in America to an operating table in Korea, where it helped save a wounded corporal. Murrow concluded with a direct appeal to listeners: "What about *your* blood? Can you spare a pint?" A mini-telethon of sorts followed, prompting a half-million blood donations across the country.[72]

The firing of the man who had directed the war garnered almost as much attention as the war itself. After profiling Douglas MacArthur in its debut broadcast, *Hear It Now* periodically touched on the growing controversy regarding the general's leadership and public pronouncements, but never at any length. That changed after President Truman relieved MacArthur of his duties on April 11, 1951. *Hear It Now* dedicated more than half of that week's program to the "supercharged atmosphere of pressure and conflict" in Washington and the outrage among Republicans and much of the nation's press. Murrow observed that all the "heat and passion appear to have caused some to forget that the war also goes on," adding, "If we are to do our duty over the long haul, we may need more stability and less hysteria and blind partisanship than we have displayed during the past three days."[73]

MacArthur delivered his famous "old soldiers never die" speech to Congress the following week. Again, *Hear It Now* devoted the brunt of its Friday night program to MacArthur, with sound from that day's ticker-tape parade in New York and lengthy excerpts of the nationally broadcast speech along with stories about the reaction to it. "Subways in New York were deserted, stores in Chicago empty, taxis in St. Louis parked and listening," said Murrow of the speech. "Had it been for just MacArthur the victorious warrior, all this would have been his due.... But returning as a displaced commander, he stood there before Congress and the nation a symbol of their own dissidence and disunity." The next week, Murrow reported that "invective and accusations [still] ran heavy," even though the nation confronted "decisions as important as any we have ever had to face." *Hear It Now* responded with what Murrow described as an attempt "to present the issues in this very serious debate": a long, sober overview of Truman and MacArthur's contrasting positions regarding Korea and the Far East. Congressional testimony highlighted the substantial concerns about MacArthur's views. By the end of May, *Hear It Now* noted that interest in MacArthur seemed to be waning, and partisan passions were displaced into the annual congressional baseball game.[74]

The MacArthur controversy in many ways paralleled what was termed the "Great Debate" over committing U.S. forces to Europe under General Dwight Eisenhower's command. President Truman asserted that he had the authority to make that commitment; Republicans insisted that he required congressional approval. The debate also pitted isolationist Republicans such as Herbert Hoover and Robert Taft against the party's internationalists such as Thomas Dewey. *Hear It Now* devoted substantial coverage to the controversy, and again Murrow lamented the angry divisions that had resulted. "No one can say how much damage the months of debate and confusion have caused in Europe among our allies, three thousand miles nearer the threat of Russia," he said after the Senate finally reached a compromise. He had been an internationalist from at least the time he covered the London Blitz, but he muted those sympathies on *Hear It Now*.[75]

The closest thing to an exception came in Murrow's comments upon the death of Senator Arthur Vandenberg, a onetime Republican isolationist who had become an eloquent advocate of a bipartisan, internationalist foreign policy. Vandenberg died at the peak of the MacArthur firestorm. "We are now divided—bitterly, hysterically," said Murrow. Vandenberg "would have gloried in this controversy, and he would have steadied it," confident that "the little men of loud voice and small faith, those who consult partisan rather than national destiny, will yield to the collective changing judgment of the American people."[76]

Foreign-policy debates played out against the specter of atomic warfare. *Hear It Now* reported on atom-bomb tests in Nevada and on a new warplane, the B-36, that could drop the bomb if necessary. The program recorded a B-36 crew's exultation upon successfully completing an arduous practice mission. The following week, Murrow somberly announced that most of the crew who had been heard in the previous program had been killed during a follow-up exercise.[77]

Hear It Now also reported on unmanned weaponry. In what was billed as "one of the most unusual recordings ever broadcast," the program aired a V-2 rocket test. The rocket's radio signal rose in pitch as the V-2 ascended into space and then dropped as the rocket decelerated and plummeted back to earth. Two weeks later, the program featured sounds of another test. "We believe you are listening to the prototype of a deadly weapon . . . the most modern military music," said Murrow. The eerie electronic drone of the weapon's sensors was heard as Murrow described it creating "its own type of glorious crescendo . . . and perhaps producing as its supreme triumph in its explosion the agonized cry of the hurt human being." The drone then broke up into static before abruptly ending in silence.[78]

Despite that story's ominous tone, Murrow was no cold-war pacifist, which the future head of the U.S. Information Agency made clear in a *Hear It Now* report on the Voice of America. In a report reminiscent of NBC's "Malice in Wonderland," which had aired the previous year, Murrow said that more "boldness and imagination" was needed to counter the "ridiculous lies" of Soviet propaganda and win the fight "to capture, or rather to liberate, men's minds." Otherwise there was "the prospect of inevitable collision, with all its consequences."[79]

Eventually, Murrow himself would collide with a senator accusing him of having "'consciously served the communist cause.'"[80] The impending confrontation with Joseph McCarthy was foreshadowed on *Hear It Now*. To be sure, Murrow's public comments concerning the Wisconsin senator in 1950 and 1951 were mild compared with what would follow some years later on television. They also were mild compared with what others in the news media said at the time. The print journalists Drew Pearson, Joseph Alsop, and Stewart Alsop were among those highly critical of McCarthy, as were radio commentators such as Elmer Davis and Martin Agronsky. One of McCarthy's chief foes in his home state was the *Capital Times* in Madison, Wisconsin; its editor, William Evjue, appeared on the NBC documentary series *Yesterday, Today, and Tomorrow* in 1951 to denounce the senator. However, Evjue complained afterward that NBC had not informed him that it would give McCarthy time in the program to reply to his criticisms. In addition, Evjue said that NBC had "mutilated" his comments by editing them down to leave only "a weak and somewhat incoherent thing that I certainly would not sponsor or authorize had I known that deletions were to be made." Moreover, the recording engineers somehow had made it sound "as though I was talking from the bottom of a barrel."[81]

Hear It Now's treatment of McCarthyism again tended toward caution, with an occasional jibe aimed in the general direction of the senator. The series premiere reported on subversion charges against Anna Rosenberg, in line for a top Pentagon post. Her successful defense prompted Murrow to declare that "the character assassin had missed." The following week, Don Hollenbeck, in his next-to-last appearance on the series, reported on a physical altercation between McCarthy and Drew Pearson. Hollenbeck said only that the McCarthy-Pearson brawl had done little to address the nation's problems.[82]

As for McCarthy himself, he was heard on the program excoriating the "crimson, motley crowd that has been selling our nation out all over the world to international communism." He called the MacArthur firing "high treason" and the Democrats "the party of betrayal," adding that the Truman administration was "preparing for another planned Pearl Harbor."[83] Finally, he launched a highly publicized attack on Defense Secretary George Marshall, the architect of the Marshall Plan and a symbol of the bipartisan internationalism of the prior decade. That prompted a Murrow response the next day on *Hear It Now*.

He began by alluding to a speech by the United Nations ambassador Warren Austin on the need for the "unity of free men who will not be divided." In a slow, chilly, ironic tone, Murrow continued: "In Washington, the cause of truth and free men who will not be divided was being served on the floor of the Senate by Joseph McCarthy of Wisconsin. . . . It wasn't until late Thursday afternoon that Senator McCarthy's colleagues gave him the floor and heard his *exposé*." Excerpts of McCarthy's Senate speech followed: Marshall was Stalin's stooge, he was to blame for Yalta and China and Korea, and so forth.

"These are merely a few of the highlights from Senator McCarthy's sixty-thousand word story of George Catlett Marshall," said Murrow, before suddenly raising his voice in pitch and volume: "*Compiled,* as he calls it, from the pens and lips of sources friendly to [Marshall]. There is *much more* that these sources have written and spoken that is *not* in the record compiled by the senator from Wisconsin." Then, "to keep the record straight," Murrow quoted effusive praise for Marshall from Winston Churchill, Admiral William Leahy, General Hap Arnold, and others before concluding:

> It is the same man whom Senator McCarthy accuses of being a leader in what he calls "a conspiracy so immense, an infamy so black, as to dwarf any in the previous history of man." And it is the same George Catlett Marshall of whom

the great Republican statesman, the late Henry L. Stimson, said at the end of the last war, "I have had considerable experience with men in government. General Marshall has given me a new gauge of what such service should be. The destiny of America at the most critical time of its national existence has been in the hands of a great and good citizen. Let no man forget it." Great and good citizen—or arch-conspirator—Secretary of Defense George Marshall is still in harness as a servant of his government.

Murrow's comments on Marshall and McCarthy aired on June 15, 1951. That same broadcast, Murrow announced that *Hear It Now* was going on summer hiatus. Before ending (for the only time during the series) with his signature, "Good night, and good luck," he promised that they would return "fortified by travel, research, and study."[84] In fact, arrangements already were being made for *Hear It Now*'s demise.

In a June 20 memo, the CBS program director Hubbell Robinson wrote to William Paley and Frank Stanton that he had been "discussing the termination of *Hear It Now*" with Murrow and Friendly, who had told him that "they wanted to spend the next six weeks studying and researching the best way to handle television news." Robinson attached a note that Friendly had sent him the previous day:

> Ed and I have begun spending many long hours on TV. The more we talk the more convinced I am that television news can never be just a translation of radio news into a medium of pictures. I think we must concern ourselves with an entirely *new concept*. . . . I think we are in the position newspapers were in before there were newspapers. I think we are where radio was before there were radio programs. We cannot merely copy or translate. We have to create. With a medium to challenge the imagination, it is time we started to stagger it.[85]

Friendly elaborated in a piece for *Variety* the following month. He declared that it was time to reinvent the documentary for the new medium. "Television will command the biggest audiences in the history of communications and of show business," he wrote. "It will enable the American people to be the best informed people in the world, or the worst, depending on how well we make use of it."[86]

If Friendly was itching for the opportunity to use television (as he had been even before arriving at CBS), Murrow remained ambivalent at best. In 1949, he had raised fears about attractive personalities drawing "huge television audiences regardless of the violence that may be done to truth or objectivity" and coverage consisting of "bathing girls on surf boards." As *Hear*

It Now began airing the next year, Murrow wrote to a CBS correspondent that he hoped "neither one of us has to try to make a living in television."[87] As late as September 1951—a month after *Billboard* reported that *See It Now* would debut that fall and the "radio counterpart of the video show will be killed"—Murrow was still telling affiliates that "management hasn't yet decided whether *Hear It Now* is coming back." Meanwhile, he reiterated his concerns about television to a New York newspaper along with his belief (if not hope) that "the premiere of 'See It Now' is still a long way off."[88]

Regardless, *See It Now* debuted that November with the iconic image of the Golden Gate Bridge and Brooklyn Bridge appearing live side by side. If that reflected Friendly's "entirely new concept," he and Murrow also would recycle several ideas from *Hear It Now*, including profiles of the troops in Korea, the biography of a pint of blood, and Carl Sandburg reading from *The People, Yes*. Later there would be the classic reports on the red-baiting targets Milo Radulovich and Annie Lee Moss as well as the climactic showdown with McCarthy. Finally would come the loss of sponsors, the clashes with CBS management, and the speech from Murrow lambasting TV's masters for using it to "distract, delude, amuse, and insulate."[89] By 1961, he would be gone from CBS.

"There are new and great possibilities in TV, but I still have the feeling I'm its prisoner and am getting pushed around by it," said Murrow at the time of *See It Now*'s premiere. "In radio I have control."[90]

* * *

One might view *Hear It Now* as little more than a trial run for the television series or as an anachronism even in its own time—"rather like building the best gas lamp at the turn of the century when most people were rewiring their homes for electricity," as one of Murrow's biographers put it. By the time the series left the air, ratings for network radio programs were plummeting as the networks shifted the brunt of their energies toward television.[91]

Yet evening radio audiences remained sizable (as shown by *Hear It Now*'s debut reaching ten million listeners on 173 stations); as of 1950, they still were larger than those for television. CBS continued to create significant radio documentaries in 1951 and 1952, such as *The Nation's Nightmare*, a six-part investigation of organized crime narrated by Bill Downs, in addition to the new version of *The People Act* that was produced by Irving Gitlin. The impact of *Hear It Now* was significant enough to mobilize mass blood donations and win a Peabody Award.[92] Whereas Murrow and Friendly's initial collaboration on the record album *I Can Hear It Now* had used old radio clips and sounded like what one historian calls "a valedictory tribute for a medium on the road to second-class status,"[93] their radio series was marked by a rich eclecticism and innovativeness that represented the most creative use of the medium that Murrow would ever achieve.

Murrow did continue his nightly radio newscast and used it to launch what a biographer has called "trial balloons for television," trying out controversial ideas before putting them on *See It Now*.[94] He also lent his voice to CBS radio documentaries later in the decade such as *Who Killed Michael Farmer?* However, the newscasts did not employ the elaborate production techniques that Friendly brought to *Hear It Now*, and Murrow did not write or report the documentaries, which would embarrass him when a program on prostitution that he had narrated provoked angry controversy. With *Hear It Now*, Murrow in fact did "have control," allowing him to expand the voices and sounds on radio and realize his stated desire to make the medium more "adult and intelligent."[95] If TV's *See It Now* would exemplify what has been described as "a rare excitement, an appetite to tackle every subject of interest under the sun," it was following in the footsteps of its radio predecessor.[96]

Hear It Now also foreshadowed the future of long-form radio journalism, much as Norman Corwin's *One World Flight* in its own way had done. In the 1970s, developers of the new public-radio newsmagazine *All Things Considered* would consciously avoid replicating what one called the "steady, authoritative, [and] a bit pompous" style of Murrow's news commentaries. Instead, they would feature highly produced stories that would be "the radio equivalent of a television news report, in which 'sound' assumed the function of 'pictures.'"[97] Thanks largely to Friendly, *Hear It Now* had helped pioneer that form two decades previously—a form in which, as one historian notes, "the real reporter was the tape recorder gathering reality sound, to which narration as needed could be added."[98]

Although the radio series was a departure for Murrow in some ways, his editorial stance on the program hewed to that for which he was already known. "He is at heart a moralist, troubled by the series of dull thuds which pass for civilization nowadays," wrote one observer before *Hear It Now* began airing. "Still, he won't preach, because he has vowed to be an objective newsman."[99] That was ironic, given Murrow's letter to his parents during the war in which he claimed to see himself as a sort of preacher with radio as his pulpit.[100] Still, for Murrow and other journalists in the postwar age, moralism and objectivity were not seen as mutually exclusive. The media historians Michael Schudson and Susan Tifft have written that "objectivity was universally acknowledged to be the spine of the journalist's moral code" at that time.[101] That was demonstrated by *Hear It Now*. It signified that the dramatized documentary featuring actors and sound effects and authored by idealistic writers bent on social change was passé. In its place was a series that stressed in its weekly introduction that "all the voices and sounds you will hear are *real*." Objective fact and straightforward reportage underscored by recorded sound and actualities became the norm and would serve as the template for news documentary on American television.[102]

Similarly, the series provided additional proof that the shift in documentary from One World liberalism to liberal anticommunism had taken firm hold by 1951. Murrow strongly supported American troops fighting a vilified enemy in Korea as well as the battle to win hearts and minds elsewhere in the world. He also scolded if not outright preached at those who did not fulfill their duties at home or who put partisan interests ahead of the national interest. There were gleams of liberal sympathies when he defended George Marshall and Anna Rosenberg against red-baiting, but he did not go so far as to criticize Joseph McCarthy directly. A 1953 *New Yorker* profile of Murrow said that McCarthy actually had commended the CBS journalist for his fairness.[103] That continuing impulse toward carefulness and impartiality frustrated those who wanted the journalist to take a bolder stand not only in his reporting but also on such issues as the network loyalty oath and the blacklist. Murrow's muted response may have partially reflected the fact that by the time he left radio for television, he had hit "a low point in his life, his health, his outlook and output," as one of his biographers has written.[104]

By the end of 1953, when the *New Yorker* profile appeared, Murrow seemed reinvigorated, at least as far as McCarthyism was concerned. *See It Now* had begun adding what Fred Friendly later described as the "missing ingredients" of "conviction, controversy, and a point of view."[105] The Milo Radulovich report already had aired; the McCarthy program soon would follow. For Murrow's biographer A. M. Sperber, that marked a decisive move away from Murrow's previous neutrality toward "adversary programming."[106] However, the contrast between the *See It Now* of 1954 and the *Hear It Now* of 1951 was not as pronounced as it might have seemed. Whatever McCarthy may have once said about Murrow's fairness, the journalist did manage to make his true feelings about the senator known on the air. Murrow's defense of George Marshall on *Hear It Now* featured what the scholar Thomas Rosteck has called "the sarcastic and satiric tenor" that would later characterize the *See It Now* report on the senator, with Murrow ironically commenting on how McCarthy's "*exposé*" of Marshall had served "the cause of truth and free men who will not be divided."[107]

Beyond that, *See It Now*'s McCarthy program made the same appeal that Murrow had regularly made on *Hear It Now*: not to "walk in fear, one of another," not to "be driven by fear into an age of unreason," not to imagine that our ancestors had "feared to write, to speak, to associate, [or] to defend causes which were for the moment unpopular." It was, argues Rosteck, an ideological appeal that celebrated the pluralistic values favored by liberal anticommunists and that stressed consensus over conflict. Yet it also was rooted in a faith consistent with that expressed by other radio documentarians such as Norman Corwin and Robert Lewis Shayon. As Murrow summarized it in *Hear It Now*, "We believe that the human ear is capable of great understanding."[108] That served as a fitting epitaph for the heyday of audio documentary on American network radio.

7

Lose No Hope

In assessing the postwar documentary's legacy, it should be asked whether Murrow was naive when he asserted in 1951 that the ear is capable of great understanding. It was, after all, a time when demagoguery was on the rise and liberal attempts at reform were in retreat simultaneous with "the final demise of a national radio service which had dominated the American scene for a quarter of a century," as one historian has put it.[1] Even during the more hopeful preceding years, when Robert Heller was speaking of the radio documentary as representing "a profound revolution" and a "virtual Utopia," the documentaries' true impact still could be questioned.[2] An analysis of the era's programs in the context of previous scholarship on radio, documentary, and journalism helps clarify the sorts of "understandings" that they promoted.

Form and Address

A tenet of journalism studies and of media studies generally is that the form of news evolves over time in response to shifts in the political, economic, technological, and cultural environment and that those changes are not preordained or invariably in the name of progress.[3] Such was the case with the radio documentary between 1945 and 1951. According to the historian A. William Bluem, it moved in fits and starts from "dramatic restatement of fact to drama made *with* fact," as new recording technology and the networks' abandonment of their recording ban encouraged the production of actuality-based programs in place of the dramatizations that had previously existed.

The radio actuality programs in turn were joined with the visual techniques that had been established by movie newsreels in providing the model for early television news documentaries such as Murrow and Friendly's *See It Now*.[4]

There was nothing inevitable about that shift, though. In response to an early CBS actuality documentary in 1947, *Variety* complained that the wire recorder "does not add to reality on radio" and that any quality actor and director "could be more effective in driving home the truth." The following year, Robert Heller excitedly foresaw video documentaries as encompassing "infinite combinations of actuality, of films, of cartoons, of animation, of studio dramatizations." Three years after that, in preparing *See It Now*, Fred Friendly wrote urgently of the need for "an entirely *new concept*" in television documentary that did not "merely copy or translate" what had existed in radio, although it turned out that Friendly and Murrow's series borrowed heavily from their radio prototype *Hear It Now*.[5] Documentary is fluid by nature; its creators may assume a number of potential roles (such as "Reporter," "Advocate," "Poet," and "Guerilla," as Erik Barnouw describes them), while again negotiating a balance between what has been called "the claim to truthfulness and the need to select and represent the reality one wants to share."[6] In the words of a British scholar looking specifically at audio documentary, the form presents a "fascinating paradox" in that it "offers authenticity, but it also denotes artifice."[7] So it was, for example, that *Living 1948* billed itself as a "drama document."

As the forms of news and documentary change, so too does the manner in which they conceive of and address their public. For example, Michael Schudson has traced how coverage of the president's annual State of the Union address evolved to grant journalists increasing power to explain and analyze the speech for a public that was implicitly "ill-equipped to sort out for itself the meaning of events." Likewise, Kevin Barnhurst and John Nerone have analyzed changes in newspaper design from the colonial era to the 1990s and argued that the twentieth-century newspaper assumed "a stance of objectivity and expertise" that privileged journalists as voices of authority at the expense of other voices.[8]

Although the present study covers only a six-year period, something similar occurred in the postwar documentary's shift in weight from "drama" to "document." Dramatized programs such as Robert Lewis Shayon's *The Eagle's Brood* did not claim to be objective; Shayon was a dramatist as opposed to a journalist, and drama is free to choose sides, as had been the case with the radio plays of the war years. In contrast, Murrow was a journalist who avoided explicit advocacy except when it seemed obvious that only one side was legitimate, as with the American troops at war in Korea. The use of recorded actuality in programs such as *Hear It Now,* as opposed to the actors, music, and sound effects of an *Eagle's Brood,* underscored an emphasis on objective fact, or as Robert Heller put it, "the cold reality of the world outside" the radio studios.[9] Recordings by themselves did not preclude advocacy; Norman Corwin's *One World Flight* argued eloquently against the escalation of the cold war (again, Corwin came from the world of drama rather than journalism). Still, actualities promoted a "realist ethos" that in many ways paralleled the introduction of photojournalism in the print news media and that similarly enhanced the journalist's authority as an objective renderer of reality.[10]

The trend toward realism signaled a change in the documentaries' relationship to what the historians Michele Hilmes and Susan J. Douglas have called the "imagined community" of American listeners during radio's heyday. As Douglas describes it, listeners were united not only by "hearing the same thing at the same time" but also by engaging "in the same cognitive and emotional work: to create a mental representation of a speaker, a news event, a story."[11] This "dimensional listening" was key to radio's power, as suggested by Murrow when he declared on *Hear It Now* that "occasionally one word or one sound is worth a dozen pictures."[12]

Dramatized and actuality forms of documentary both could promote dimensional listening, as with *The Eagle Brood's* evocation of the Back of the Yards neighborhood in Chicago or *Hear It Now*'s vivid depiction of a pint of blood's journey to Korea. However, the dramatized programs drew heavily upon the idealism and social consciousness of the Depression and war years. They aspired to a documentary model characterized by what has been called "socially useful storytelling." The result, according to the radio historian Bruce Lenthall, was "a deliberate emphasis on innovative artistry and, crucially, using that artistry to convey socially charged messages" by taking full advantage of radio as a medium that "leapt across space and time."[13]

Fred Friendly similarly saw radio as a "'magic carpet'" that could go anywhere, "'maybe even the moon and the stars.'" Significantly, though, he spoke in terms of the new audio recorder "'taking a man with a story'" to anywhere he wished, including beyond the earth's atmosphere (which seemed a realistic possibility by the end of the war).[14] Not for Friendly were ABC's 1947 depiction of Jiminy and Grumpy aboard a magic rocket ship to the future or the dramatized reenactments of history offered by *CBS Is There.* Although he was hardly above tweaking reality to suit his needs, his orientation was toward the reporter uncovering fact on location by means of the recorder and later the camera—toward a "Life Magazine of the Air."[15]

The realist turn in audio documentary matched the changing political climate. Typically, dramatized documentaries invited listeners to imagine a better world, and actuality documentaries invited listeners to view the world as it really was, more intractable than perfectible. That was not universally true of all the era's programs; *Communism—U.S. Brand* and *Living*'s "Malice in Wonderland" series were docudramas that purported to arm citizens with knowledge about the communist threat, whereas *The People Act* used recordings to encourage citizens to join together in improving their communities. Nevertheless, there was a growing sense by 1951 that the idealistic docudramas of past years had trafficked in fantasy that was outmoded if not naive. Corwin's assertion soon after World War II that "if oneness of the world is a dream, then we are proud to call ourselves dreamers" was no longer in fashion.[16] Actuality programs presenting a harder-nosed realism and stressing the need for vigilance in dangerous and uncertain times were now the norm, as with *Hear It Now* presenting the "brutal naked sounds" of the war in Korea.

According to William Bluem, that model of journalist-centered documentary stressing "precision and impartiality of description, with emphasis on the detached and dispassionate in techniques of presentation," carried over to television, although "poetic" forms of documentary focusing on "the 'world of imagination'" and "universal themes of life and humanity" endured in the new medium. Still, Thomas Rosteck in his study of *See It Now* argues that a hard distinction cannot be drawn between the journalistic and poetic or between the objective and the subjective. He sees documentaries as "simultaneously records of the world beyond and arguments about that world." Even though cloaked in the conventions of objectivity, programs such as *See It Now*'s reports on McCarthyism took clear sides while expressing the "ideological motifs" of mainstream liberalism: "an emphasis upon the common man, the valuation of controversy, and a defense of the pluralistic society."[17]

Such motifs were characteristic of postwar audio documentaries in general, whether in the hands of a dramatist such as Corwin or a journalist such as Murrow. Michele Hilmes has written of the "fundamental tensions" between "utopian predictions of radio's unifying power" (including its "utopian discourse of uplift and education") and "the dystopian possibilities that radio had to be kept from unleashing"—base consumerism, seemingly low or vulgar forms of popular culture, and other perceived threats to "progressive white Americanism."[18]

Postwar documentary expressed radio's utopian side. It explicitly appealed to public service and sought to transcend commercialism, with the vast majority of the programs airing without sponsorship to comparatively small, elite audiences. It articulated the mores of "progressive white Americanism," which by the end of the war included an acceptance of racial and religious difference. Still, it was produced virtually exclusively by whites for a white-majority listenership. It also was produced by men with only a few exceptions, and although it sometimes spoke respectfully of women, it often, consciously or not, was patronizing toward them. In many cases, it vigorously advocated social change, but within limits; widespread systemic overhaul was not offered as a possibility. It embraced values common in American journalism, as the sociologist Herbert Gans has described them: altruistic democracy and moderatism (tolerating disagreement and dissent but not demagoguery), responsible capitalism and individualism (rejecting collectivism while celebrating the "common man"), and ethnocentrism along with small-town pastoralism (promoting classically American virtues, especially as the cold war intensified). Such values were as much conservative as liberal in spirit. As frequently has been the case with American radio and with network news documentary throughout their histories, the postwar audio documentary criticized the status quo at the same time as it upheld it.[19]

Regulation and Competition

Previous studies have stressed the influence of regulation and commercial competition on American television news documentary. Examples include the programs of the early 1960s that were produced partly as a response to fallout from the quiz-show scandals of the previous decade and to pressure from regulators who saw TV as a "vast wasteland." Muckraking exposés in the late 1960s and early 1970s benefitted from the cover provided by government regulatory protection. With deregulation and rising financial pressures in the 1980s, network television documentary went into decline.[20]

Such pressures similarly influenced the rise and fall of the postwar audio documentary. There is no "smoking gun" within the available archival evidence to prove that the networks produced documentaries in direct response to the specter of tougher regulation that the FCC's Blue Book represented. However, there is substantial circumstantial evidence to suggest that the FCC report and pressures from media reformers did play a role. Not least was the formation of the CBS Documentary Unit only six months after the Blue Book's issuance and William Paley's assertion soon afterward to the National Association of Broadcasters that they needed to "earn and hold public confidence" via "new and sparkling ideas in the presentation of educational, documentary, and controversial issues." Press commentaries about the documentaries referred to the FCC report, with Jack Gould praising one program by saying that if the networks produced more shows like it, "'blue books' criticizing radio would be a thing of the past." A journal article coauthored by Charles Siepmann (who had a vested interest in the Blue Book) declared that such programs were "a partial answer to the insistence of the Federal Communications Commission, enunciated in the so-called 'Blue Book,' that radio devote more and better time to programs in the public interest."[21]

Competition among the networks also provided an early impetus to the documentaries, with CBS elevating its sustaining programming to differentiate itself from NBC, ABC investing in ambitious public-service productions to garner attention at a time when it had not much else going for it, and NBC belatedly seeking to catch up with its rivals. Eventually the documentaries would fall victim to what a historian has called "a familiar cycle in the history of American broadcast programming: innovation, imitation, saturation."[22] CBS (which by 1948 was aggressively seeking to overtake NBC as the top-rated network) decided that there were "'too many' news documentaries on the air" and that "it had squeezed the radio documentary for all the value it could get."[23] Although CBS did vigorously promote *Hear It Now* during its short run, the series quickly shifted to television, as did the majority of the network's attention and resources. The same was true of NBC and ABC, with Robert Saudek of the latter network leaving commercial broadcasting for the Ford Foundation.

The changing political climate contributed to the documentaries' decline as well. The Blue Book ultimately had a limited impact on radio due to broadcasters' vociferous opposition to the report and a new Republican majority in Congress that was far less sympathetic to regulatory reform than New Deal liberals had been.[24] The move among liberals toward anticommunism, the intensification of the cold war (along with the eruption of actual war in Korea), and the blacklisting of many of top radio documentarians ended the genre's heyday.

Power and Poetry

The discussion so far has lent considerable support to David Paul Nord's assertion that "the 'consciousness' embedded in the language of journalism is the product of large institutions" and "the exercise of power." Nonetheless, it would be wrong to succumb to what James Carey termed an "anti-Whig interpretation of the press," reflecting a "contemptuous view from the academy toward journalism." James Ettema and Theodore Glasser similarly have questioned those who assert "in a tone of ironic knowingness that the media are inevitably the means to hegemonic power rather than democratic empowerment."[25]

Contrary to such views and assertions, the best of postwar audio documentary is an exemplar of commercial broadcasting networks—the very embodiment of the mainstream, corporate media—granting free airtime to the theme of democratic empowerment. The programs may not always have represented a literal "exercise in poetry and utopian politics," as Carey described journalism in its ideal form (although some documentary scripts actually were written in verse, as was the case with some of Corwin's work as well as that of *Living*'s Lou Hazam).[26] Still, they did represent the poetic strain of documentary, stressing the "'world of imagination'" and "universal themes of life and humanity."[27] Those included what Richard Rorty described as "utopian dreams [of] an ideally decent and civilized society," or as Robert Lewis Shayon put it, "an American democracy ever-expanding rather than contracting the flow of its benefits to all the people."[28] The same was true even of the harder-edged and more "realistic" brand of news documentary represented by Murrow and Friendly. One scholar has argued that Murrow's peroration about Joseph McCarthy on *See It Now* (whose roots again can be traced back to radio) "scan[ned] like free verse" while echoing "the common sense of Thomas Paine and the rhetorical rhythms of Abraham Lincoln," suggesting that Murrow never had completely abandoned the notion of broadcasting as a kind of secular pulpit.[29]

Therein lie the "moral and political ambiguities" of journalism and documentary that Carey said historians should seek to uncover.[30] The postwar audio documentaries were enabled and constrained by institutional self-interest. At the same time, they reflected the hopes and aspirations of not only their writers but also the moguls in charge of the networks that aired them. As Shayon put it, for a time after the war, "the smell of wartime clung to the men in the grey flannel suits, and in their civvies, they continued to honor the national purpose."[31] Scholars should continue to be sensitive to such historical moments when flux and ambiguity in the institutional and cultural environment make room for innovative, public-spirited work, including that produced by the mainstream media, which often are seen as having been traditionally inhospitable to such work. We ought to maintain a keen focus on the "exercise of power" during such moments, as Nord argued; at the same time, we should avoid the pitfalls of "anti-Whigism," or what Michael Schudson has described as "declinism": the notion that the press inexorably grows worse over time.[32]

We also should be mindful of the broader lessons of the postwar audio documentary. For those, we could do worse than to turn to Albert Einstein. In 1946, he participated in Robert Lewis Shayon's *Operation Crossroads*, which called for international control of the atomic bomb. Shayon had a very specific role in mind for Einstein, telling him, "I want you to say that human nature can be changed, and that we must change it in order to survive in the atomic age." In short, Shayon wanted Einstein's scientific imprimatur on the idea that humankind could be perfected, in accordance with the fondest liberal hopes of the day.

"When you talk of changing human nature, are you talking of historical time, geological time, or mathematical time?" Einstein asked. When a puzzled Shayon asked Einstein what he meant, the physicist explained that certainly human nature could be changed in mathematical time (viewed in terms of infinity), and possibly it could be changed in geological time (viewed in terms of millions of years). In terms of historical time, spanning recorded human experience, Einstein said that it was clear that human nature could *not* be changed: "But it is precisely because we cannot change human nature in historical time that we must do this [impose arms control]. When a river overflows its banks and destroys lives and property, do we sit by the banks of the river and weep because we cannot change the course of the river? No. We build dams to contain it. In the same manner, we must build institutions to restrain the fears and suspicions and greed which move people and their rulers."[33]

Einstein's words from 1946 apply equally well to today's media. The postwar documentary developed partly as a response to public and regulatory calls for better radio. Structural reform helping to ensure a more truly competitive media marketplace by containing the worst excesses of commercialism and conglomeration can help nurture more vigorous and more public-minded journalism and documentary.[34]

The following year, in 1947, Ruth Ashton approached Einstein to participate in *The Sunny Side of the Atom* regarding atomic energy. This time, Einstein turned down the invitation. As she recalled, "[H]e didn't feel that a radio program could have any effect. He didn't feel you could really tell the story you needed to tell about the importance of the atomic age unless, as he said, you told it [directly to people on] the street corners."[35]

Again, Einstein's words relate to today—not in the sense that the media can have no positive impact on society but that the media alone cannot be expected to effect all necessary change for the better. To accomplish that change, as Archibald MacLeish said at the conclusion of *Operation Crossroads*, citizens must be continually "talking and listening and thinking."[36]

Contemporary audio documentary produced by professionals and by citizens themselves can play a modest role in fostering that conversation. Prisoners, teenagers, and others have told their own stories, bridging divisions of race, class, and age. The Internet allows easy distribution of those stories and grants them a permanence that such work had not previously enjoyed when it was broadcast only once over the radio. The stories provide listeners with direct access to other voices and take advantage of the "dimensional listening" that audio offers. "Lacking earlids, we are defenseless, vulnerable to ambush," the documentarian Jay Allison has said. "Invisibility is our friend. Prejudice is suspended while the listener is blind, only listening." When such documentaries are aired on American public radio, they enable that medium to move nearer to its founding "utopian ideal that we use these airwaves to share our stories as we try to understand each other better."[37] They thus help accomplish what has been called the "most urgent" value that journalism and documentary can promote: a sense of human solidarity providing "insight into difference—whether among neighbors or nations—and a search for common ground."[38]

* * *

It is too easy to mythologize the pioneers of radio or to wax nostalgic for radio's supposed Golden Age.[39] Edward R. Murrow himself betrayed nostalgia in his famous 1958 speech lambasting television before the Radio-Television News Directors Association when he said that "that most satisfying and rewarding instrument" of radio ought to "go backward" to a time when it was unsullied by commercialism and was "proud [and] alert"—that is, when it seemingly was poised to show the postwar world exactly what it could do to further the hard-won peace.[40] As we have seen, that era's legacy is more ambiguous than romanticized nostalgia is apt to recognize. Still, our efforts to puncture or demythologize nostalgia should not go so far as to overlook a historical moment when notions of collective conscience and responsibility still could be taken seriously and radio and mass media still could be seen as means of upholding such values. Not least, we should remember the postwar audio documentary for its implicit admonition that the fault for falling short of our noblest aspirations has never rested anywhere else than with ourselves.

The final word can be given to Norman Corwin, who was still alive and working as he passed his hundredth birthday. He had lived long enough to see works such as *One World Flight* find an appreciative audience among the new generation of audio documentarians. "Many of us were not yet conceived when Norman Corwin's words reverberated through the air, carrying the weight of their time, and hope for the future," Jay Allison has written. "But they have echoed down to us. We who choose to work in radio are certain to encounter them along our way, and when we do, our very sentences may be altered."[41]

Corwin remained an unapologetic liberal who, when interviewed in 2006, was not sanguine about the state of the world: "I must say that my optimism is strained every time that I turn on the tube or read the morning paper."[42] Issues he had confronted six decades previously—warfare, religious strife, poverty, resentment and suspicion toward America, questions regarding individual loyalty and patriotism—were still in the news each day.

Nevertheless, Corwin still felt enough confidence to quote from the end of the speech he had delivered upon his return to America in 1946, fresh from his journey around a war-shattered globe with a microphone and a primitive wire recorder: "I have lost no hope. I believe that ultimately we will find unity and brotherhood in this world, but that the quest will go on through terrible trials and agonies, until a true democracy, not merely a lip-service democracy, is achieved for the entire world."[43]

Notes

Introduction: Utopian Dreams

1. Qtd. in Philip Seib, *Broadcasts from the Blitz: How Edward R. Murrow Helped Lead America into War* (Washington, D.C.: Potomac, 2006), 147–48.
2. Edward R. Murrow, *In Search of Light: The Broadcasts of Edward R. Murrow, 1938–1961,* ed. Edward Bliss Jr. (New York: Knopf, 1967), 94.
3. A. M. Sperber, *Murrow: His Life and Times* (1986; reprint, New York: Fordham University Press, 1998), 254. See also Stanley Cloud and Lynne Olson, *The Murrow Boys* (Boston: Houghton Mifflin, 1996).
4. Robert Lewis Shayon, *Odyssey in Prime Time* (Philadelphia: Waymark, 2001), 12.
5. See William N. Robson, "Open Letter on Race Hatred," in *Radio Drama in Action,* ed. Erik Barnouw (New York: Rinehart, 1945), 57–77.
6. Qtd. in Sperber, *Murrow,* 254.
7. Robert P. Heller, "The Dynamic Documentary," in *Radio and Television Writing,* rev. ed., ed. Max Wylie (New York: Rinehart, 1952), 383. Heller's essay was first published in 1949, prior to his listing in *Red Channels* and subsequent departure from CBS.
8. Christopher H. Sterling and John M. Kittross, *Stay Tuned: A Concise History of American Broadcasting* (Belmont, Calif.: Wadsworth, 1978), 248.
9. A. William Bluem, *Documentary in American Television* (New York: Hastings, 1965), 71.
10. Ibid., 72.
11. Michael Curtin, *Redeeming the Wasteland: Television Documentary and Cold War Politics* (New Brunswick, N.J.: Rutgers University Press, 1995), 7.
12. Ibid.; Chad Raphael, *Investigated Reporting: Muckrakers, Regulators, and the*

Struggle over Television Documentary (Urbana: University of Illinois Press, 2005). See also Thomas Rosteck, *See It Now Confronts McCarthyism* (Tuscaloosa: University of Alabama Press, 1994); Thomas A. Mascaro, "The Chilling Effects of Politics: CBS News and Documentaries during the Fin-Syn Debate in the Reagan Years," *American Journalism* 22.4 (2005): 69–97.

13. Edward Bliss Jr., *Now the News: The Story of Broadcast Journalism* (New York: Columbia University Press, 1991); Lawrence W. Lichty, "Documentary Programs on U.S. Radio," in *Museum of Broadcast Communications Encyclopedia of Radio*, vol. 1, ed. Christopher H. Sterling (New York: Fitzroy Dearborn, 2004), 474.

14. Christopher H. Sterling and Michael C. Keith, "Where Have All the Historians Gone? A Challenge to Researchers," *Journal of Broadcasting and Electronic Media* 50.2 (2006): 345–57.

15. "Radio Documentaries Take Listeners into Dark Corners," *Nieman Reports* 55.3 (Fall 2001): 13–14. See also John Biewen and Alexa Dilworth, eds., *Reality Radio: Telling True Stories in Sound* (Chapel Hill: University of North Carolina Press, 2010).

16. Daniel Makagon and Mark Neumann, *Recording Culture: Audio Documentary and the Ethnographic Experience* (Los Angeles: Sage, 2009). For examples of the work of the Lomaxes and Schwartz, see Alan Lomax, "Mister Ledford and the TVA," in *Radio Drama in Action*, ed. Erik Barnouw (New York: Rinehart, 1945), 48–58; and Tony Schwartz, *Songs of My City: The Stories, Music, and Sounds of the People of New York* (New York: Folkways Records, 1956).

17. See, for example, Michele Hilmes and Jason Loviglio, eds., *Radio Reader: Essays in the Cultural History of Radio* (New York: Routledge, 2002); Michael C. Keith, ed., *Radio Cultures: The Sound Medium in American Life* (New York: Peter Lang, 2008); Elizabeth Fones-Wolf, *Waves of Opposition: Labor and the Struggle for Democratic Radio* (Urbana: University of Illinois Press, 2006).

18. Susan J. Douglas, *Listening In: Radio and the American Imagination* (New York: Times Books), 24; Michele Hilmes, *Radio Voices: American Broadcasting, 1922–1952* (Minneapolis: University of Minnesota Press, 1997), 11–33. See also Benedict Anderson, *Imagined Communities: Reflections on the Origins and Spread of Nationalism* (New York: Verso, 1983); William Stott, *Documentary Expression and Thirties America* (1973; reprint, Chicago: University of Chicago Press, 1986), 80–91; Bruce Lenthall, *Radio's America: The Great Depression and the Rise of Modern Culture* (Chicago: University of Chicago Press, 2007).

19. Susan Merrill Squier, "Communities of the Air: Introducing the Radio World," in *Communities of the Air: Radio Century, Radio Culture*, ed. Susan Merrill Squier (Durham, N.C.: Duke University Press, 2003), 6.

20. Bluem, *Documentary in American Television*, 68.

21. Patricia Aufderheide, *Documentary Film: A Very Short Introduction* (New York: Oxford University Press, 2007), 127.

22. James W. Carey, "The Problem of Journalism History" (1974), in *James Carey:*

A Critical Reader, ed. Eve Stryker Munson and Catherine A. Warren (Minneapolis: University of Minnesota Press, 1997), 88–90.

23. James W. Carey, "The Press and the Public Discourse," *Center Magazine* 20.2 (March/April 1987): 14.

24. David Paul Nord, "A Plea for *Journalism* History," *Journalism History* 15.1 (1988): 10–11, 14.

25. Richard Rorty, *Achieving Our Country: Leftist Thought in Twentieth-Century America* (Cambridge, Mass.: Harvard University Press, 1998), 43, 106.

26. See Michael Denning, *The Cultural Front: The Laboring of American Culture in the Twentieth Century* (London: Verso, 1997); Judith E. Smith, "Radio's 'Cultural Front,' 1938–1948," in *Radio Reader: Essays in the Cultural History of Radio,* ed. Michele Hilmes and Jason Loviglio (New York: Routledge, 2002), 209–30.

27. Lenthall, *Radio's America,* 192–93. See also Howard Blue, *Words at War: World War II Era Radio Drama and the Postwar Broadcasting Industry Blacklist* (Lanham, Md.: Scarecrow, 2002).

28. *Public Service Responsibility of Broadcast Licensees* (Washington, D.C.: Federal Communications Commission, 1946).

29. Norman Corwin interviewed by Douglas Bell, *Years of the Electric Ear* (Metuchen, N.J.: Directors Guild of America/Scarecrow, 1994), 135–36.

30. Robert Lewis Shayon, "Murrow's Lost Fight," *Saturday Review,* May 22, 1965, 94.

31. Sperber, *Murrow,* 704.

32. See Gary Edgerton, "The Murrow Legend as Metaphor: The Creation, Appropriation, and Usefulness of Edward R. Murrow's Life Story," *Journal of American Studies* 15.1 (1992): 75–91. See also W. Joseph Campbell, *Getting It Wrong: Ten of the Greatest Misreported Stories in American Journalism* (Berkeley: University of California Press, 2010), 45–67.

33. Thomas Doherty, *Cold War, Cool Medium: Television, McCarthyism, and American Culture* (New York: Columbia University Press, 2003), 162–63. For one example of a broadcasting history that disputes Murrow's importance, see Douglas Gomery, *A History of Broadcasting in the United States* (Malden, Mass.: Blackwell, 2008).

34. Shayon was the source regarding Murrow's alleged exhortation, which was quoted in A. M. Sperber's biography *Murrow* (Sperber interviewed Shayon for the book). The book claims that Murrow met Shayon and others, including Charles Collingwood, at a hotel bar in Paris in April 1945. In his own memoirs published several years after the Murrow biography, Shayon said that the meeting actually took place in London with Murrow, Collingwood, and Douglas Edwards, all of CBS. Shayon wrote of that meeting only that Murrow had "laid out his intentions" for postwar radio, leaving Shayon "in a mood of optimism and anticipation." Shayon's 1945 diary confirms some of those details—the meeting was in fact in London, and Edwards was present—but says it happened in June rather than April and that CBS's Larry LeSueur was there instead of Collingwood. (According to the diary, Shayon had just dined

with Collingwood in Paris before traveling to London.) The diary records Murrow talking about how Winston Churchill "loved the power and glory too much to retire as he should have and written his memoirs," whereas Clement Attlee was "dull and unimposing." (Attlee was about to be replace Churchill as British prime minister.) Bernard Montgomery was "a bad general" and a "prima donna" but "good at inspiring his men." General Dwight Eisenhower was "simple in tastes—not brilliant" but still praised for his "wonderful work" during the war. The conversation also touched on the "confusing and muddling" leadership of the U.S. State Department and "the need for enlightening the American people" toward a greater "friendship with Russia." See Sperber, *Murrow*, 254, 704; Shayon, *Odyssey in Prime Time*, 90; Robert Lewis Shayon diary entry, June 14, 1945, Robert Lewis Shayon Papers (Box 9, Folder 52, Diary 3), Howard Gotlieb Archival Research Center, Boston University.

35. David Everitt, *A Shadow of Red: Communism and the Blacklist in Radio and Television* (Chicago: Ivan R. Dee, 2007), 194.

36. Michele Hilmes, *Only Connect: A Cultural History of Broadcasting in the United States* (Belmont, Calif.: Wadsworth, 2002), 133. See also Barbara Dianne Savage, *Broadcasting Freedom: Radio, War, and the Politics of Race, 1938–1948* (Chapel Hill, N.C.: University of North Carolina Press, 1999).

37. Paddy Scannell, "'The Stuff of Radio': Developments in Radio Features and Documentaries before the War," in *Documentary and the Mass Media*, ed. John Corner (London: Edward Arnold, 1986), 26. For a critique of the public-affairs documentary subgenre, see Aufderheide, *Documentary Film*, 56–65.

38. See Robert W. McChesney, *Telecommunications, Mass Media, and Democracy: The Battle for the Control of U.S. Broadcasting, 1928–1935* (New York: Oxford University Press, 1993); Gerd Horten, *Radio Goes to War: The Cultural Politics of Propaganda during World War II* (Berkeley: University of California Press, 2002); Elizabeth Fones-Wolf, *Selling Free Enterprise: The Business Assault on Labor and Liberalism, 1945–1960* (Urbana: University of Illinois Press, 1994); Victor Pickard, "Media Democracy Deferred: The Postwar Settlement for U.S. Communications, 1945–1949," Ph.D. dissertation, University of Illinois at Urbana-Champaign, 2008.

39. Shayon, *Odyssey in Prime Time*, 91.

40. See Aufderheide, *Documentary Film*, 1–55.

41. Stott, *Documentary Expression and Thirties America*, 73, 314.

42. See Daniel C. Hallin, "The Passing of the 'High Modernism' of American Journalism," *Journal of Communication* 42.3 (1992): 14–25; Kevin G. Barnhurst and John Nerone, *The Form of News: A History* (New York: Guilford, 2001); John C. Nerone, ed., *Last Rights: Revisiting Four Theories of the Press* (Urbana: University of Illinois Press, 1995).

43. "What Should Be the Criteria for Broadcasting in the Public Interest, Convenience, and Necessity?" in *Education on the Air: The Seventeenth Yearbook of the Institute for Education by Radio*, ed. O. Joe Olson (Columbus: Ohio State University, 1947), 69.

44. James W. Carey, "'Putting the World at Peril': A Conversation with James W. Carey," in *James Carey: A Critical Reader*, ed. Eve Stryker Munson and Catherine A. Warren (Minneapolis: University of Minnesota Press, 1997), 109.

45. Shayon, *Odyssey in Prime Time*, 145.

Chapter 1: A Higher Destiny

1. Clayton R. Koppes, "The Social Destiny of the Radio: Hope and Disillusionment in the 1920s," *South Atlantic Quarterly* 68.3 (1969): 364, 370. See also Michele Hilmes, *Radio Voices: American Broadcasting, 1922–1952* (Minneapolis: University of Minnesota Press, 1997), 11–33.

2. Robert W. McChesney, *Telecommunications, Mass Media, and Democracy: The Battle for Control of U.S. Broadcasting, 1928–1935* (New York: Oxford University Press, 1993).

3. John Wallace, writing in 1928, qtd. in Koppes, "Social Destiny of the Radio," 372.

4. Lawrence W. Lichty and Thomas W. Bohn, "Radio's *March of Time:* Dramatized News," *Journalism Quarterly* 51 (1974): 460. A recording of *The March of Time*'s radio premiere on March 6, 1931, is available at the Paley Center for Media in New York City.

5. Raymond Fielding, *The March of Time: 1935–1951* (New York: Oxford University Press, 1978), 17–18.

6. Erik Barnouw, *Documentary: A History of the Non-Fiction Film*, 2d rev. ed. (New York: Oxford University Press, 1993), 122.

7. Fielding, *March of Time*, 18.

8. Howard Blue, *Words at War: World War II Era Radio Drama and the Postwar Broadcasting Industry Blacklist* (Lanham, Md.: Scarecrow, 2002), 162–63.

9. Erik Barnouw, *Media Marathon: A Twentieth-Century Memoir* (Durham, N.C.: Duke University Press, 1996), 74–78. *Cavalcade of America* remained on radio until 1953. See also John Dunning, *On the Air: The Encyclopedia of Old-Time Radio* (New York: Oxford University Press, 1998), 141–42; Judith E. Smith, "Radio's 'Cultural Front,' 1938–1948," in *Radio Reader: Essays in the Cultural History of Radio*, ed. Michele Hilmes and Jason Loviglio (New York: Routledge, 2002), 214–15.

10. William Stott, *Documentary Expression and Thirties America* (1973; reprint, Chicago: University of Chicago Press, 1986), xi, 21.

11. Ibid., 9; Barnouw, *Documentary*, 85. For overviews of Grierson's work, see Barnouw, *Documentary*, 84–100; Patricia Aufderheide, *Documentary Film: A Very Short Introduction* (New York: Oxford University Press, 2007), 32–38.

12. A. William Bluem, *Documentary in American Television* (New York: Hastings, 1965), 49–51. See also Aufderheide, *Documentary Film*, 66–67; Stott, *Documentary Expression*, 23–24.

13. Stott, *Documentary Expression*, 90.

14. Bruce Lenthall, *Radio's America: The Great Depression and the Rise of Modern Culture* (Chicago: University of Chicago Press, 2007), 4.

15. Qtd. in Blue, *Words at War*, 17.

16. Erik Barnouw, *The Golden Web: A History of Broadcasting in the United States, 1933–1953* (New York: Oxford University Press, 1968), 55–73.

17. Ibid., 67. See also Archibald MacLeish, "The Fall of the City," in *Radio's Best Plays*, ed. Joseph Liss (New York: Greenberg, 1947), 3–31.

18. Archibald MacLeish, "Of Poetic Drama," *New York Times*, April 4, 1937, sec. 10, 2.

19. See Edward Bliss Jr., *Now the News: The Story of Broadcast Journalism* (New York: Columbia University Press, 1991), 39–44; Barnouw, *Golden Web*, 18–22.

20. Bliss, *Now the News*, 25–30.

21. Ibid., 25–26, 97, 106–8.

22. Qtd. in Stott, *Documentary Expression*, 89; Alexander Kendrick, *Prime Time: The Life of Edward R. Murrow* (Boston: Little, Brown, 1969), 195. For an overview of news coverage of the buildup to the war and the war's outbreak, see Bliss, *Now the News*, 81–142.

23. Qtd. in Philip Seib, *Broadcasts from the Blitz: How Edward R. Murrow Helped Lead America into War* (Washington, D.C.: Potomac, 2006), 148–49; Bliss, *Now the News*, 134.

24. Qtd. in Bliss, *Now the News*, 139.

25. See, for example, Douglas Gomery, *A History of Broadcasting in the United States* (Malden, Mass.: Blackwell, 2008), ix–xii, 46.

26. Edward R. Murrow, *In Search of Light: The Broadcasts of Edward R. Murrow, 1938–1961*, ed. Edward Bliss Jr. (New York: Knopf, 1967), 73, 85, 94.

27. Barnouw, *Documentary*, 139.

28. See ibid., 155–63; Bluem, *Documentary in American Television*, 54–56.

29. Erik Barnouw, ed., *Radio Drama in Action* (New York: Rinehart, 1945), 386.

30. Norman Rosten, "Concerning the Red Army"; Morton Wishengrad, "The Battle of the Warsaw Ghetto"; Arthur Laurents, "The Last Day of the War"; all in *Radio Drama in Action*, ed. Erik Barnouw (New York: Rinehart, 1945), 31–45, 91–110, 165–80.

31. Arch Oboler, "The House I Live In," in *Radio Drama in Action*, ed. Erik Barnouw (New York: Rinehart, 1945), 385–97. Oboler's radio play should not be confused with the 1945 Oscar-winning short film of the same title that starred Frank Sinatra and that similarly centered around the song.

32. Roi Ottley, "The Negro Domestic," in *Radio Drama in Action*, ed. Erik Barnouw (New York: Rinehart, 1945), 356.

33. Harry Kleiner, "Japanese-Americans," in *Radio Drama in Action*, ed. Erik Barnouw (New York: Rinehart, 1945), 237.

34. William N. Robson, "Open Letter on Race Hatred," in *Radio Drama in Action*, ed. Erik Barnouw (New York: Rinehart, 1945), 60.

35. Ibid., 75. See also Barnouw, *Golden Web*, 181–83; "Outspoken Broadcast," *Time*, August 9, 1943, 62.

36. Michele Hilmes, *Only Connect: A Cultural History of Broadcasting in the United States* (Belmont, Calif.: Wadsworth, 2002), 133.

37. Langston Hughes to Erik Barnouw, March 27, 1945, Erik Barnouw Papers (Box 38), Rare Book and Manuscript Library, Columbia University, New York. See also Barbara Dianne Savage, *Broadcasting Freedom: Radio, War, and the Politics of Race, 1938–1948* (Chapel Hill, N.C.: University of North Carolina Press, 1999); Smith, "Radio's 'Cultural Front.'"

38. "Outspoken Broadcast," 62; John K. Hutchens, "Radio Notebook," *New York Times,* August 1, 1943, 7.

39. Norman Corwin, "The Radio," in *While You Were Gone: A Report on Wartime Life in the United States,* ed. Jack Goodman (New York: Simon and Schuster, 1946), 377.

40. Mike Conway, *The Origins of Television News in America: The Visualizers of CBS in the 1940s* (New York: Peter Lang, 2009), 158.

41. Columbia Broadcasting System, *Columbia Broadcasting System, Inc., Annual Report for the Fiscal Year Ended December 29, 1945* (New York: Columbia Broadcasting System, 1946), 4–5. The poll that CBS quoted was conducted by Paul Lazarsfeld and would be published in book form later in 1946; see Columbia University Bureau of Applied Social Research, *The People Look at Radio* (Chapel Hill, N.C.: University of North Carolina Press, 1946).

42. Bliss, *Now the News,* 186.

43. Columbia Broadcasting System, *Columbia Broadcasting System Report* (1945), 16.

44. National Broadcasting Company, *Annual Review: The National Broadcasting Company, 1945–46* (New York: National Broadcasting Company, 1946), 16.

45. Ibid., 17.

46. Bliss, *Now the News,* 180–82.

47. Columbia Broadcasting System, *Columbia Broadcasting System, Inc., Annual Report for the Fiscal Year Ended December 28, 1946* (New York: Columbia Broadcasting System, 1947), 18–19.

48. National Broadcasting Company, *Annual Review,* 20–21.

49. See Smith, "Radio's 'Cultural Front,'" 222–24; Ottley, "Negro Domestic"; Bliss, *Now the News,* 181.

50. "*Variety*'s Capsule Appraisal of Radio's Know-It-Alls," *Variety,* July 25, 1945, 26, 30. See also Bliss, *Now the News,* 54–65, 140–41; Irving E. Fang, *Those Radio Commentators!* (Ames: Iowa State University Press, 1977).

51. Conway, *Origins of Television News,* 160.

52. Columbia Broadcasting System, *Columbia Broadcasting System Annual Report* (1945), 10.

53. National Broadcasting Company, *Annual Review,* 31.

54. Quincy Howe, "Policing the Commentator: A News Analysis," *The Atlantic*, November 1943, 46–49. See also Bryce Oliver, "Thought Control—American Style," *New Republic*, January 13, 1947, 12–13; J. Fred MacDonald, *Don't Touch That Dial! Radio Programming in American Life, 1920–1960* (Chicago: Nelson-Hall, 1979), 315–18.

55. "Radio, Too, Can Talk Back," *Variety*, October 24, 1945, 35.

56. See Saul Carson, "Gabbers vs. 'Dies' Fight Starts 'Thought Control' Mobilization," *Variety*, October 24, 1945, 35, 46; "Protests Snowballing vs. House Gag Bill; May Wipe Out Committee Itself," *Variety*, December 5, 1945, 26. See also "Quincy Howe, Broadway Legiter, in 'Dies' Quiz; Walton Menaced," *Variety*, October 31, 1945, 37, 44.

57. Joseph C. Goulden, *The Best Years: 1945–1950* (New York: Atheneum, 1976), 5.

58. Emil Corwin and Alan Reitman, "Is Radio Going Liberal?" *New Republic*, February 12, 1945, 220.

59. Goulden, *Best Years*, 166.

60. Barnouw, *Radio Drama in Action*, vii.

61. Eleanor E. Timberg, "The Mythology of Broadcasting," *Antioch Review* 6.3 (September 1946): 367.

62. Paul A. Porter, "Radio Must Grow Up," *American Magazine* 140.10 (October 1945): 24–25.

63. See *Public Service Responsibility of Broadcast Licensees* (Washington, D.C.: Federal Communications Commission, 1946); Robert D. Leigh, ed., *A Free and Responsible Press* (Chicago: University of Chicago Press, 1947); Victor Pickard, "Reopening the Postwar Settlement for U.S. Media: The Origins and Implications of the Social Contract between Media, the State, and the Polity," *Communication, Culture, and Critique* 3.2 (2010): 170–89.

64. Victor Pickard, "Media Democracy Deferred: The Postwar Settlement for U.S. Communications, 1945–1949," Ph.D. dissertation, University of Illinois at Urbana-Champaign, 2008, 26. See also Elizabeth Fones-Wolf, *Waves of Opposition: Labor and the Struggle for Democratic Radio* (Urbana: University of Illinois Press, 2006), 125–61.

65. Corwin, "The Radio," 374–75.

Chapter 2: One World

1. Biographical information is from R. LeRoy Bannerman, *On a Note of Triumph: Norman Corwin and the Golden Years of Radio* (New York: Carol Publishing Group, 1986).

2. Qtd. in Michael C. Keith, "Norman Corwin: Words in Flight," *Journal of Radio Studies* 5.2 (1998): 59.

3. Norman Corwin, telephone interview with the author, February 12, 2006.

4. Bannerman, *On a Note of Triumph*, 39.

5. Ibid., 7.

6. Norman Corwin, *Thirteen by Corwin* (New York: Henry Holt, 1942), 78.

7. See Bannerman, *On a Note of Triumph*, 73–143; Norman Corwin, *Untitled and Other Radio Dramas* (New York: Henry Holt, 1947); Norman Corwin, "This Is War!," in *Education on the Air: Thirteenth Yearbook of the Institute for Education by Radio*, ed. Josephine H. McClatchy (Columbus: Ohio State University, 1942), 87–92; A. M. Sperber, *Murrow: His Life and Times* (1986; reprint, New York: Fordham University Press, 1998), 212–14.

8. Gerd Horten, *Radio Goes to War: The Cultural Politics of Propaganda during World War II* (Berkeley: University of California Press, 2002).

9. Corwin, "This Is War!," 89–90.

10. Corwin, *Untitled*, 441. A film about Corwin's V.E.-Day program, *A Note of Triumph: The Golden Age of Norman Corwin*, won the 2006 Academy Award for Best Documentary Short.

11. "More by Corwin," *Time*, May 28, 1945, 66; "Best Busts," *Time*, August 27, 1945, 58.

12. Philip Hamburger, "Profiles: The Odyssey of the Oblong Blur," *New Yorker*, April 5, 1947, 42.

13. "More by Corwin," 66. See also Jack Gould, "A Minority Report," *New York Times*, May 20, 1945, sec. 2, 5.

14. Corwin, *Untitled*, 220. Significantly, Corwin in *On a Note of Triumph* contrasted his idealized "little guy" with the fascist "little man who could be relied on to take orders." See ibid., 460–61.

15. See William Stott, *Documentary Expression and Thirties America* (1973; reprint, Chicago: University of Chicago Press, 1986); Susan Faludi, *Stiffed: The Betrayal of the American Man* (New York: Morrow, 1999), 16–24; Susan J. Douglas, *Listening In: Radio and the American Imagination* (New York: Times Books, 1999), 194.

16. Corwin, *Untitled*, 536; Keith, "Norman Corwin," 59–60. See also Stephen Vincent Benét, *We Stand United and Other Radio Scripts* (New York: Farrar and Rinehart, 1945); Archibald MacLeish, "The Fall of the City," in *Radio's Best Plays*, ed. Joseph Liss (New York: Greenberg, 1947), 3–31. Others whom Corwin esteemed included the journalist Heywood Broun and fellow radio writers such as Lucille Fletcher and Paul Rhymer (respectively best known for the suspense thriller *Sorry, Wrong Number*, and the lighthearted serial *Vic and Sade*); see Keith, "Norman Corwin."

17. "Bill of Rights," *Variety*, December 17, 1941, 44.

18. Richard Glatzer, "*Meet John Doe*: An End to Social Mythmaking," in *Meet John Doe*, ed. Charles Wolfe (New Brunswick, N.J.: Rutgers University Press, 1989), 245; Gilbert Seldes, *The Great Audience* (New York: Viking, 1950), 120.

19. Henry A. Wallace, *The Century of the Common Man* (New York: Reynal and Hitchcock, 1943), 19–20.

20. Norman D. Markowitz, *The Rise and Fall of the People's Century: Henry A. Wallace and American Liberalism, 1941–1948* (New York: Free Press, 1973), 48.

21. Qtd. in Michael C. Keith and Mary Ann Watson, eds., *Norman Corwin's One World Flight: The Lost Journal of Radio's Greatest Writer* (New York: Continuum, 2009), 192–94. Corwin was an official of the Independent Citizens Committee of the Arts, Sciences, and Professions, members of which would back Wallace for president in 1948. See "Corwin Radio Chrm. for Arts, Sciences Comm.," *Variety*, May 2, 1945, 42; David Everitt, *A Shadow of Red: Communism and the Blacklist in Radio and Television* (Chicago: Ivan R. Dee, 2007), 24–25.

22. Corwin, *Untitled*, 428, 462, 471.

23. Markowitz, *Rise and Fall of the People's Century*, 48.

24. Corwin, *Untitled*, ix, 479, 484. See also Norman Corwin, "What Peace and Whose Radio?" *Variety*, January 9, 1946, 114, 122.

25. Biographical information is from Ellsworth Barnard, *Wendell Willkie: Fighter for Freedom* (Marquette: Northern Michigan University Press, 1966); and Philip Beidler, "Remembering Wendell Willkie's One World," *Canadian Review of American Studies* 24.2 (1994): 87–104.

26. Wendell Willkie, *One World* (1943; reprint, Urbana: University of Illinois Press, 1966), 191, 202. Willkie expressed similar sentiments on race relations in a spoken postscript to the 1943 CBS program *Open Letter on Race Hatred*; see William N. Robson, "Open Letter on Race Hatred," in *Radio Drama in Action*, ed. Erik Barnouw (New York: Rinehart, 1945), 75–77.

27. Beidler, "Remembering Wendell Willkie's One World," 103.

28. "Prizes for Corwin," *Time*, March 4, 1946, 62; "CBS Plans Global Show Cued by Corwin Jaunt in Memory of Willkie," *Variety*, May 8, 1946, 26; Keith and Watson, *Norman Corwin's One World Flight*, 1–3.

29. See Louis Liebovich, *The Press and the Origins of the Cold War* (New York: Praeger, 1988), 125–38.

30. Norman Corwin speech, February 18, 1946, Norman Corwin Collection (box 40, folder 437), Special Collections, Thousand Oaks Library, Thousand Oaks, Calif. (hereafter Corwin Collection); Norman Corwin speech, March 31, 1946, Corwin Collection (box 40, folder 450).

31. See Paul Boyer, *By the Bomb's Early Light: American Thought and Culture at the Dawn of the Atomic Age* (New York: Pantheon, 1985), 65–75, 203–10; Spencer R. Weart, *Nuclear Fear: A History of Images* (Cambridge, Mass.: Harvard University Press, 1988), 116–17; John Hersey, *Hiroshima* (New York: Knopf, 1946); Patrick B. Sharp, "From Yellow Peril to Japanese Wasteland: John Hersey's 'Hiroshima,'" *Twentieth Century Literature* 46.4 (2000): 434–52.

32. *Hiroshima*, part 1, originally broadcast September 9, 1946, recording available at Paley Center for Media, Beverly Hills. See also Saul Carson, "ABC's Sock but Humble Airing of 'Hiroshima' Gives Radio New Stature," *Variety*, September 11, 1946, 34.

33. See Robert Lewis Shayon, *Odyssey in Prime Time* (Philadelphia: Waymark, 2001), 91–100. A further discussion of *Operation Crossroads* follows in chapter 3.

34. *Unhappy Birthday*, originally broadcast August 6, 1946, recording available at

Paley Center for Media, Beverly Hills. See also "Unhappy Birthday," *Variety,* August 14, 1946, 29, 33.

35. Qtd. in Bannerman, *On a Note of Triumph,* 178.

36. Corwin telephone interview with the author, February 12, 2006.

37. Ibid.; Corwin, *Untitled,* 339–406.

38. "Disks Catch On," *Business Week,* June 21, 1947, 69.

39. Other reasons given for the recording ban include a stigma attached to recordings due to FCC requirements that they be identified as such on the air, concerns about audio quality, and pressure from musicians' unions. An exception to the ban was made for sound effects. See Barnouw, *Golden Web,* 109–10, 216–19; Edward Bliss Jr., *Now the News: The Story of Broadcast Journalism* (New York: Columbia University Press, 1991), 36–38, 157–58.

40. Alan Lomax, "Mister Ledford and the TVA," in *Radio Drama in Action,* ed. Erik Barnouw (New York: Rinehart, 1945), 47–58. See also Daniel Makagon and Mark Neumann, *Recording Culture: Audio Documentary and the Ethnographic Experience* (Los Angeles: Sage, 2009), 3–9.

41. See Paddy Scannell, "'The Stuff of Radio': Developments in Radio Features and Documentaries before the War," in *Documentary and the Mass Media,* ed. John Corner (London: Edward Arnold, 1986), 19–20.

42. "Documentary Programs: A Demonstration and Discussion," in *Education on the Air: Twelfth Yearbook of the Institute for Education by Radio,* ed. Josephine H. McClatchy (Columbus: Ohio State University, 1941), 245–49. See also Christopher H. Sterling and John M. Kittross, *Stay Tuned: A Concise History of American Broadcasting* (Belmont, Calif.: Wadsworth, 1978), 206–7.

43. Barnouw, *Golden Web,* 245; Sterling and Kittross, *Stay Tuned,* 251; "Recorded Broadcasts Better Than Radio, Insures Perfection—Jolson," *Variety,* September 25, 1946, 1; "Corwin Trek Cues CBS Switch on Anti-Wax Ban," *Variety,* October 30, 1946, 35, 46; "Webs in Persecution Complex," *Variety,* December 11, 1946, 29; "Large Crew Races Clock for Preem of Corwin 'One World Flight' Jan. 14," *Variety,* January 1, 1947, 23. See also "Disks Catch On"; "Transcription Boom," *Newsweek,* January 19, 1948, 58, 61; "Tape for the Networks," *Newsweek,* May 3, 1948, 52.

44. Qtd. in Ralph Engelman, *Friendlyvision: Fred Friendly and the Rise and Fall of Television Journalism* (New York: Columbia University Press, 2009), 46.

45. See "Documentary Reporting," in *Education on the Air: Fourteenth Yearbook of the Institute for Education by Radio,* ed. Josephine H. McClatchy (Columbus: Ohio State University, 1943), 172–75; Edward R. Murrow, *In Search of Light: The Broadcasts of Edward R. Murrow, 1938–1961,* ed. Edward Bliss Jr. (New York: Knopf, 1967), 83–85; Bliss, *Now the News,* 200–201; Sterling and Kittross, *Stay Tuned,* 206–7.

46. Keith and Watson, *Norman Corwin's One World Flight,* 5.

47. Ibid., 4–5; letters and memos in Corwin Collection (box 40, folder 442; box 41, folder 451); Norman Corwin to Bill Costello, June 14, 1946, in *Norman Corwin's Letters,* ed. A. J. Langguth (New York: Barricade, 1994), 94–95; "One World Flight," *All*

Things Considered, National Public Radio, August 13, 1999; accessed August 6, 2010, http://www.npr.org/programs/lnfsound/stories/990813.stories.html. George Polk would be murdered in 1948 while covering the Greek Civil War; see Kati Marton, *The Polk Conspiracy* (New York: Farrar, Straus, and Giroux, 1990).

48. *One World Flight* scrapbook, Corwin Collection. Although the letter was dated the day of Corwin's departure, he would not receive it until a couple of weeks later when he arrived in Stockholm. See Keith and Watson, *Norman Corwin's One World Flight*, 40–41.

49. *Corwin Departure for One World Flight*, originally broadcast June 15, 1946, recording available at Paley Center for Media, Beverly Hills.

50. "CBS Picks Brains, Ups Budget, to Lift Show Values," *Variety*, January 2, 1946, 23; "CBS Plans Global Show."

51. CBS memo from A. B. Chamberlain to Davidson Taylor, April 26, 1946, Corwin Collection (box 41, folder 451).

52. Qtd. in Keith and Watson, *Norman Corwin's One World Flight*, 5, 22–24, 46–47; "One World Flight," *All Things Considered*. See also Bannerman, *On a Note of Triumph*, 179–80.

53. Keith and Watson, *Norman Corwin's One World Flight*, 43.

54. Ibid., 61–62, 88.

55. Corwin and Bland did manage to gain a private audience (sans recorder) with the Pope at the Vatican. See ibid., 105–8.

56. Norman Corwin to Erik Barnouw, February 18, 1968, in *Norman Corwin's Letters*, ed. A. J. Langguth (New York: Barricade, 1994), 271–72.

57. Corwin telephone interview with the author, February 12, 2006; Keith and Watson, *Norman Corwin's One World Flight*, 148–52.

58. Keith and Watson, *Norman Corwin's One World Flight*, 160–61.

59. Norman Corwin to William Paley, August 7, 1946, Corwin Collection (box 36, folder 416).

60. News releases, office communications, and script drafts in Corwin Collection (box 40, folder 433, and box 39, folder 428); Lee Bland, "Lee Bland Tips Wire Recorder Technique for Quality Programs," *Billboard*, April 5, 1947, 3, 14; "Large Crew Races Clock"; Bliss, *Now the News*, 200–201.

61. Corwin telephone interview with the author, February 12, 2006.

62. News releases and series outline in Corwin Collection (box 40, folders 433 and 446); "Large Crew Races Clock," 23.

63. Series outline in Corwin Collection (box 40, folder 446).

64. "Corwin Trek Cues CBS Switch," 46. See also Bliss, *Now the News*, 200–201.

65. *One World Flight*, January 14, 1947. Recordings of the series are available in the Corwin Collection and at the Paley Center for Media in New York and Beverly Hills. They also are available from Jerry Haendiges Productions (http://otrsite.com). For a published script of the premiere, see Norman Corwin, "One World Flight," *Hollywood Quarterly* 2.3 (1947): 280–88.

66. *One World Flight*, January 21, 1947.
67. *One World Flight*, January 28, 1947, and February 4, 1947.
68. *One World Flight*, February 4, 1947.
69. Corwin, *Untitled*, 399–400.
70. *One World Flight*, February 11, 1947.
71. *One World Flight*, February 18, 1947, and February 25, 1947.
72. *One World Flight*, March 4, 1947.
73. *One World Flight*, March 11, 1947.
74. *One World Flight*, March 18, 1947.
75. *One World Flight*, March 25, 1947, and April 1, 1947.
76. Norman Corwin, "One World Revisited," *Common Ground* 7.2 (1947): 3–17.
77. *One World Flight*, April 8, 1947.
78. Jerome Lawrence, "A New Radio Form," *Hollywood Quarterly* 2.3 (1947): 280–81; Jerry Franken, "Superb Corwin Preem Reveals Bigotry, Ignorance Rampant," *Billboard*, January 25, 1947, 8; "One World Flight," *Variety*, January 22, 1947, 32; Seymour Peck, "Corwin's 'One World Flight' Brings Hope for Peace," *PM*, January 15, 1947, 20.

79. Jack Gould, "One World Flight," *New York Times*, February 2, 1947, sec. 2, 11; "The World and Norman Corwin," *Time*, January 27, 1947, 65; "Corwin's World," *Newsweek*, February 10, 1947, 58; Lou Frankel, "In One Ear," *The Nation*, February 15, 1947, 187.

80. CBS mail analysis in Corwin Collection (box 40, folder 447); Corwin, *Untitled*, 535.

81. "Corwin-Hope Boxscore," *Variety*, January 29, 1947, 25; Saul Carson, "Whose World Flight?" *New Republic*, April 7, 1947, 41–42. According to *Variety*, Bob Hope was either the number-one or number-two program in the country for every ratings period while Corwin's series was on the air, with Hope averaging a 31.4 rating. "Total network competition" (that is, *One World Flight* combined with all other network programs airing against Hope) averaged only 6.2. See "Top Fifteen and the Opposition," *Variety*, January 22, 1947, 28; February 5, 1947, 34; February 19, 1947, 25; March 5, 1947, 26; March 19, 1947, 33; April 2, 1947, 31; April 23, 1947, 24.

82. "Corwin and Luce for Com'l Series?," *Variety*, June 12, 1946, 27; "Luce Loses Corwin," *Variety*, June 26, 1946, 31.

83. "Corwin Odyssey at Half-Way Mark," *Variety*, August 14, 1946, 22, 30.

84. Corwin telephone interview with the author, February 12, 2006.

85. Corwin, "The Radio," 381. See also Norman Corwin to Bertha White Nason, January 26, 1945, in *Norman Corwin's Letters*, A. J. Langguth (New York: Barricade, 1994), 84–85.

86. "CBS Picks Brains."

87. Norman Corwin speech, March 31, 1946, in Corwin Collection (box 40, folder 450).

88. "Paley Urges Stronger NAB Standards," *Broadcasting*, October 28, 1946, 15, 101; "Corwin's Projected CBS Series Implements Paley's 'Program Primer,'" *Variety*,

November 13, 1946, 24. See also Corwin, "The Radio," 381–84; Norman Corwin to Clifford Durr, March 8, 1946, in *Norman Corwin's Letters*, ed. A. J. Langguth (New York: Barricade, 1994), 92–93; *Public Service Responsibility of Broadcast Licensees* (Washington, D.C.: Federal Communications Commission, 1946); Barnouw, *Golden Web*, 227–36; Victor Pickard, "Media Democracy Deferred: The Postwar Settlement for U.S. Communications, 1945–1949," Ph.D. dissertation, University of Illinois at Urbana-Champaign, 2008, 191–241.

89. Corwin telephone interview with the author, February 12, 2006. "My [ratings] numbers weren't *that* bad," Corwin said during the same interview. "It wasn't as though I were broadcasting from the Siberian desert!" Still, in contrast to *One World Flight*, Bob Hope's series was sponsored by Pepsodent, which in 1947 was giving away cars and other newly available consumer goods to Hope's listeners. See Carson, "Whose World Flight?"

90. "CBS 'Behind-the-Hope-Ball' Treatment on Corwin Nixes Web's Fine Words," *Variety*, December 18, 1946, 39; Bryce Oliver, "Thought Control—American Style," *New Republic*, January 13, 1947, 12–13; "CBS Fiddles as 'Liberals' Burn," *Variety*, January 29, 1947, 25. See also "In the Public Interest," *Variety*, January 8, 1947, 107; Barnard Rubin, "Broadway Beat," *Daily Worker*, January 1, 1947, 12; Frank Scully, "Gab Casualties Mounting, but Liberals Fight Back," *Variety*, January 8, 1947, 113; "Unhealthy Rumors," *Variety*, January 29, 1947, 25; Franken, "Superb Corwin Preem"; Sperber, *Murrow*, 276–77; Carson, "Whose World Flight?"

91. "CBS Fiddles as 'Liberals' Burn," 25.

92. Norman Corwin to Corwin family, August 17, 1946, in *Norman Corwin's Letters*, ed. A. J. Langguth (New York: Barricade, 1994), 97; Norman Corwin notes for news conference, January 10, 1947, in Corwin Collection (box 40, folder 438); Norman Corwin interviewed by Douglas Bell, *Years of the Electric Ear* (Metuchen, N.J.: Directors Guild of America/Scarecrow, 1994), 67.

93. "Radio Production Section Meeting," in *Education on the Air: The Seventeenth Yearbook of the Institute for Education by Radio*, ed. O. Joe Olson (Columbus: Ohio State University Press, 1947), 388. See also Sperber, *Murrow*, 298–305.

94. Corwin, *Untitled*, 401.

95. Keith and Watson, *Norman Corwin's One World Flight*, 5, 82. See also Howard Blue, *Words at War: World War II Era Radio Drama and the Postwar Broadcasting Industry Blacklist* (Lanham, Md.: Scarecrow, 2002), 21–22, 239–48.

96. Frankel, "In One Ear," 187.

97. Max Wylie, *Radio and Television Writing*, rev. ed. (New York: Rinehart, 1952), 132. See also Seldes, *Great Audience*, 119–21.

98. *One World Flight*, January 14, 1947; Corwin, "One World Flight," 283, 288. See also Jay Allison, "Radio Storytelling Builds Community On-Air and Off," *Nieman Reports* 55.3 (Fall 2001): 16–17. For his part, Corwin took exception to the argument that he dealt only in abstractions rather than recognizable individuals; see Norman

Corwin to R. LeRoy Bannerman, November 26, 1973, in *Norman Corwin's Letters,* ed. A. J. Langguth (New York: Barricade, 1994), 314–15.

99. Corwin telephone interview with the author, February 12, 2006.

100. Lawrence, "New Radio Form," 281.

101. *One World Flight,* February 11, 1947.

102. "One World Flight," *Variety,* January 22, 1947, 32; Franken, "Superb Corwin Preem," 8.

103. Lawrence, "New Radio Form," 280.

104. News release in Corwin Collection (box 40, folder 433).

105. "Probers Move In on Radio; Seek Corwin Scripts," *Variety,* April 23, 1947, 23; Corwin telephone interview with the author, February 12, 2006.

106. Norman Corwin, "The Keynote," in *Thought Control in the USA,* ed. Harold L. Salemson (New York: Garland, 1977), 34.

107. Bill Costello to Norman Corwin, June 27, 1947, Corwin Collection (box 36, folder 419). See also "3d Party Aim Laid to U.S. Communists," *New York Times,* July 22, 1947, 6; Norman Corwin to Robert Heller, July 26, 1947, in *Norman Corwin's Letters,* ed. A. J. Langguth (New York: Barricade, 1994), 101; Bannerman, *On a Note of Triumph,* 194–98. Corwin also had headed the "Mobilization against Thought Police in the USA" campaign against HUAC's investigations of radio commentators back in 1945; see Saul Carson, "Gabbers-vs.-'Dies' Fight Starts 'Thought Control' Mobilization," *Variety,* October 24, 1945, 35, 46.

108. Report on Norman Corwin, December 1, 1947, in American Business Consultants *Counterattack* Research Files (box 27, folder 14–37), Robert F. Wagner Labor Archives, Tamiment Library, New York University.

109. "CBS Fiddles as 'Liberals' Burn," 25.

Chapter 3: New and Sparkling Ideas

1. See, for example, David Halberstam, *The Powers That Be* (1979; reprint, New York: Laurel, 1986), 176–92; Sally Bedell Smith, *In All His Glory: The Life and Times of William S. Paley and the Birth of Modern Broadcasting* (New York: Touchstone, 1991), 229–73.

2. William S. Paley, *As It Happened* (Garden City, N.Y.: Doubleday, 1979), 173–74.

3. Joseph E. Persico, *Edward R. Murrow: An American Original* (New York: Laurel, 1990), 244.

4. Robert Lewis Shayon, *Odyssey in Prime Time* (Philadelphia: Waymark, 2001), 147. For an account of Paley's activities during the war, see Smith, *In All His Glory,* 203–27.

5. "CBS Picks Brains, Ups Budget, to Lift Show Values," *Variety,* January 2, 1946, 23.

6. Jack Gould, "Backward Glance," *New York Times,* December 29, 1946, 47.

7. Charles A. Siepmann, *Radio's Second Chance* (Boston: Little, Brown, 1946), 65, 73.

8. *Public Service Responsibility of Broadcast Licensees* (Washington, D.C.: Federal Communications Commission, 1946), 12, 17, 42, 55. For a comprehensive discussion of the Blue Book and the radio-reform movement from which it emerged, see Victor Pickard, "Media Democracy Deferred: The Postwar Settlement for U.S. Communications, 1945–1949," Ph.D. dissertation, University of Illinois at Urbana-Champaign, 2008.

9. "Let's Face It!" *Variety*, March 13, 1946, 35.

10. "Blue Book a Red Herring to Miller as NAB Prexy Waves Rotarian Flag," *Variety*, September 25, 1946, 42. See also Richard J. Meyer, "Reaction to 'The Blue Book,'" *Journal of Broadcasting* 6 (1962): 295–312; Michael J. Socolow, "Questioning Advertising's Influence over American Radio: The Blue Book Controversy of 1945–1947," *Journal of Radio Studies* 9.2 (2002): 282–302; Erik Barnouw, *The Golden Web: A History of Broadcasting in the United States, 1933–1953* (New York: Oxford University Press, 1968), 227 36.

11. George Rosen, "Radio Must Reform—or Else," *Variety*, October 23, 1946, 1, 90; "Paley's Primer on Programming," *Variety*, October 23, 1946, 90; Paley, *As It Happened*, 173–74. For more on CBS's previous attempts to block increased regulation, see Smith, *In All His Glory*.

12. "Documentary Unit Established by CBS Program Department to Produce Broadcasts on Major U.S. and World Issues," CBS news release, September 12, 1946, CBS News Archives Reference Library, New York (hereafter CBS News Archives); George Rosen, "CBS' 'Let the People Know': Super-Cuffos to Bump Com'ls," *Variety*, November 20, 1946, 41.

13. See Loren Ghiglione, *CBS's Don Hollenbeck: An Honest Reporter in the Age of McCarthyism* (New York: Columbia University Press, 2008); Loren Ghiglione, ed., *Radio's Revolution: Don Hollenbeck's* CBS Views the Press (Lincoln: University of Nebraska Press, 2008).

14. "Preparation of the Documentary Broadcast," in *Education on the Air: The Seventeenth Yearbook of the Institute for Education by Radio*, ed. O. Joe Olson (Columbus: Ohio State University, 1947), 377.

15. Arnold Perl, "The Empty Noose," in *Radio's Best Plays*, ed. Joseph Liss (New York: Greenberg, 1947), 122.

16. Ibid., 129–31.

17. "The Empty Noose," *Variety*, October 23, 1946, 30.

18. "CBS' 'Let the People Know,'" 41.

19. Ruth Ashton Taylor interview with Shirley Biagi for the Washington Press Club Foundation, September 11, 1991, 52; accessed August 6, 2010, http://wpcf.org/oralhistory/tay2.html. Although she could not recall her exact salary, Ashton did say that it was less than one hundred dollars a week.

20. Robert P. Heller, "Reporting by Radio," *New York Times*, October 26, 1947, 11;

Robert P. Heller, "The Dynamic Documentary," in *Radio and Television Writing*, rev. ed., ed. Max Wylie (New York: Rinehart, 1952), 383.

21. *The Land Is Bright* script, July 21, 1945, Robert Lewis Shayon Papers (box 14, folder 72), Howard Gotlieb Archival Research Center, Boston University (hereafter Shayon Papers); Robert Lewis Shayon, "Bob Shayon, Back from Europe, Tips Radio to Sell Peace as Well as Soap," *Variety*, June 27, 1945, 24.

22. *Operation Crossroads* final broadcast script, May 28, 1946, Shayon Papers (box 6, folder 34). See also Shayon, *Odyssey in Prime Time*, 91–100.

23. "60-Min. Do-Gooders Get Paley-CBS Nod," *Variety*, June 19, 1946, 24.

24. "Inside Radio," *Variety*, July 24, 1946, 46. See also "Radio's Influence on Children," *New York Times*, April 14, 1946, 55; "Justice Dept. Aide Defends Pix-Radio in Crime Inquiry," *Variety*, November 27, 1946, 1, 55; Jack Gould, "Children's Programs: Protests on Crime Shows Raise Subject Anew," *New York Times*, March 2, 1947, 75.

25. Sidney Lohman, "One Thing and Another," *New York Times*, April 14, 1946, 55; Jack Gould, "Adoption of Code in Radio Expected," *New York Times*, September 19, 1947, 46; "Paley's Primer."

26. Shayon, *Odyssey in Prime Time*, 101.

27. "Truman Endorses Delinquency Curb," *New York Times*, January 4, 1946, 20; "Clark Fears Era of Lawless Youth," *New York Times*, May 12, 1946, 15. See also Bess Furman, "Agencies Map Fight on Teen-Age Crime," *New York Times*, November 21, 1946, 36.

28. "Juvenile Delinquency Lifts Youthful Crime 100 Percent," *Life*, April 8, 1946, 92.

29. See Bess Furman, "Urges New Agency on Child Welfare," *New York Times*, December 2, 1943, 24; James Gilbert, *A Cycle of Outrage: America's Reaction to the Juvenile Delinquent in the 1950s* (New York: Oxford University Press, 1986); Sanford D. Horwitt, *Let Them Call Me Rebel: Saul Alinsky—His Life and Legacy* (New York: Knopf, 1989).

30. Saul Alinsky, "Heads I Win and Tails You Lose," undated manuscript, Shayon Papers (box 3, folder 12).

31. Horwitt, *Let Them Call Me Rebel*, 105; Saul Alinsky, *Reveille for Radicals* (Chicago: University of Chicago Press, 1946), 73.

32. Alinsky, *Reveille for Radicals*, 29; Ralph Bates, "Rhetoric for Radicals," *The Nation*, April 20, 1946, 481–82.

33. Horwitt, *Let Them Call Me Rebel*, 102–4; "Democracy in the 'Jungle,'" *New York Herald Tribune*, August 21, 1940, 20; Gretta Palmer, "Back of the Yards—Democracy with Teeth," *Reader's Digest*, March 1946, 123–26; Helena Huntington Smith, "We Did It Ourselves," *Woman's Home Companion*, May 1946, 24, 78–79; Agnes E. Meyer, "Orderly Revolution—The Back of the Yards Neighborhood Council," *Washington Post*, June 4, 1945, 9.

34. Horwitt, *Let Them Call Me Rebel*, 108.

35. "Preparation of the Documentary Broadcast," 378.

36. See Robert Lewis Shayon to Edward R. Murrow, August 26, 1946, Shayon Papers (box 3, folder 12); "CBS Man Tells of Sad Journey," *Lincoln State Journal,* March 23, 1947, n.p., Shayon Papers (box 27, folder 132); "CBS Producer Shayon on Month's Tour of Country Seeks Data for Juvenile Delinquency Documentary," CBS news release, October 10, 1946, Shayon Papers (box 115, folder 6).

37. Shayon, *Odyssey in Prime Time,* 103; Horwitt, *Let Them Call Me Rebel,* 205.

38. Robert Lewis Shayon, "Juvenile Delinquency Program," ca. 1946, Shayon Papers (box 3, folder 12).

39. Shayon, *Odyssey in Prime Time,* 101–3; Bruce Lenthall, *Radio's America: The Great Depression and the Rise of Modern Culture* (Chicago: University of Chicago Press, 2007), 178, 197.

40. Shayon, *Odyssey in Prime Time,* 103.

41. Ibid., 104; Edward R. Murrow to William S. Paley, December 10, 1946, Shayon Papers (box 3, folder 12).

42. David O. Selznick to William S. Paley, February 25, 1947, Shayon Papers (box 3, folder 12).

43. "Rising Menace of Juvenile Delinquency in U.S. Exposed by CBS in Documentary, 'Eagle's Brood,' Based on Three-Month Study," CBS news release, February 14, 1947, Shayon Papers (box 3, folder 12).

44. "CBS President Frank Stanton Wires Mayors, Washington Officials to Hear Documentary on Juvenile Delinquency," CBS news release, February 27, 1947, Shayon Papers (box 3, folder 12); "Federation of Women's Clubs in 166 Cities to Use CBS Documentary on Delinquency as Springboard for Local Action," CBS news release, February 26, 1947, CBS News Archives.

45. Shayon, *Odyssey in Prime Time,* 104–5.

46. These and subsequent quotations are taken from Robert Lewis Shayon, *The Eagles' Brood,* original broadcast recording, March 5, 1947, Radio Program Archive, University of Memphis, Tenn.; Robert Lewis Shayon, "The Eagle's Brood" (script), in *Radio and Television Writing,* rev. ed., ed. Max Wylie (New York: Rinehart, 1952), 386–406.

47. Shayon, *Odyssey in Prime Time,* 105.

48. "What Should Be the Criteria for Broadcasting in the Public Interest, Convenience, and Necessity?" in *Education on the Air: The Seventeenth Yearbook of the Institute for Education by Radio,* ed. O. Joe Olson (Columbus: Ohio State University, 1947), 84.

49. Saul Alinsky to Valentine E. Macy Jr., March 8, 1947, Industrial Areas Foundation Records (folder 173), Daley Library Special Collections, University of Illinois at Chicago.

50. Jack Gould, "'The Eagle's Brood,'" *New York Times,* March 9, 1947, 11.

51. "Between the Ears," *Time,* March 17, 1947, 93.

52. Albert N. Williams, "Eagle's Brood," *Saturday Review,* April 12, 1947, 62.

53. Lou Frankel, "In One Ear," *The Nation,* March 15, 1947, 304.

54. "CBS' 'Eagle's Brood' Flies High in Public Service," *Billboard,* March 15, 1947, 7.

55. John Crosby, "Radio in Review: 'The Eagle's Brood,'" *New York Herald Tribune,* March 11, 1947, 25.

56. George Rosen, "Juve Problem (the Nation's Canker) Unveiled by CBS in Notable Broadcast," *Variety,* March 12, 1947, 46.

57. Howard Fitzpatrick, "Radio Listening," *Boston Post,* ca. March 1947, n.p., Shayon Papers (box 27, folder 132); Seymour Peck, "'The Eagle's Brood' Gives New Importance to Radio," *PM,* March 18, 1947, 19.

58. See "'Eagle's Brood': 6.4," *Variety,* March 12, 1947, 39; C. E. Butterfield, "Radio Day by Day," *Eau Claire Leader,* March 18, 1947, n.p., Shayon Papers (box 27, folder 132); Williams, "Eagle's Brood."

59. Rudolph Elie Jr., "Reserved for Radio: Radio Double Billing Irritant at Worst for Those Who Missed Airing of 'Eagle's Brood,'" *Boston Herald,* March 8, 1947, n.p., Shayon Papers (box 115, scrapbook).

60. "Radio Production," in *Education on the Air: The Seventeenth Yearbook of the Institute for Education by Radio,* ed. O. Joe Olson (Columbus: Ohio State University, 1947), 381; Elmo C. Wilson, "The Effectiveness of Documentary Broadcasts," *Public Opinion Quarterly* 12.1 (1948): 19–29.

61. J. R. Cominsky to Frank Stanton, March 25, 1947, Shayon Papers (box 3, folder 13).

62. "CBS Documentary Unit's One-Hour Broadcast, 'A Long Life and a Merry One,' to Study Impending Crisis in U.S. Public Health," CBS news release, March 19, 1947, CBS News Archives; "Long Life and a Merry One," *Variety,* April 9, 1947, 30.

63. "CBS Psychologist at Work on 'Experiment in Living' Stresses Need for Applying Democracy in Daily Life," CBS news release, May 28, 1947, CBS News Archives.

64. Ruth Ashton Taylor interview with Shirley Biagi for the Washington Press Club Foundation, May 11, 1992, 145–46; accessed August 10, 2010, http://wpcf.org/oralhistory/tay5.html. Biographical information comes from this source.

65. See Edward Bliss Jr., *Now the News: The Story of Broadcast Journalism* (New York: Columbia University Press, 1991), 101–3.

66. Ruth Ashton Taylor, telephone interview with the author, June 30, 2008.

67. "Preparation of the Documentary Broadcast," 379.

68. Taylor interview with Biagi, September 11, 1991, 46.

69. Ibid., 46–49.

70. Ibid., 49.

71. Ibid., 50–51; Taylor telephone interview with author, June 30, 2008; Taylor interview with Biagi, May 11, 1992, 145.

72. Wilson, "Effectiveness of Documentary Broadcasts," 28.

73. These and subsequent quotations are taken from Carl Beier and Ruth Ashton, "The Sunny Side of the Atom," in *The Best One-Act Plays 1947–1948,* ed. Margaret Mayorga (New York: Dodd, Mead, 1948), 27–53.

74. "The Sunny Side of the Atom," *Variety,* July 2, 1947, 27.

75. Taylor telephone interview with author, June 30, 2008; Taylor interview with Biagi, September 11, 1991, 51.

76. "Programs in Review," *New York Times,* July 6, 1947, 51; "Documenting the Atom," *Newsweek,* July 7, 1947, 56; "Radio Also Has Its Sunny Side," *Christian Century,* July 16, 1947, 869.

77. Heller, "Reporting by Radio," 11; Heller, "Dynamic Documentary," 386.

78. Robert Lewis Taylor, "Let's Find Out," *New Yorker,* January 18, 1947, 32.

79. Ernest Dichter, *The Strategy of Desire* (Garden City, N.Y.: Doubleday, 1960), 59, 264.

80. See Vance Packard, *The Hidden Persuaders* (New York: David McKay, 1957); Barbara B. Stern, "The Importance of Being Ernest: Commemorating Dichter's Contribution to Advertising Research," *Journal of Advertising Research* 44.2 (2004): 165–69.

81. Shayon, *Odyssey in Prime Time,* 92–93. See also "CBS Psychologist Prepares Special File as Guide in Juvenile Serials," *Variety,* August 15, 1945, 22.

82. Judith Klein, "Entertainment-Education Plan Urged for Radio," *New York Herald Tribune,* June 16, 1946, sec. 2, 5.

83. Ernest Dichter, "Eagle's Brood: Juvenile Delinquency," unpublished report, ca. 1946, Shayon Papers (box 3, folder 12), 11–13; Shayon, *Odyssey in Prime Time,* 102.

84. See David L. Sills, "Stanton, Lazarsfeld, and Merton—Pioneers in Communication Research," in *American Communication Research: The Remembered History,* ed. Everette E. Dennis and Ellen Wartella (Mahwah, N.J.: Lawrence Erlbaum, 1996), 105–16; Tore Hallonquist and John Gray Peatman, "Diagnosing Your Radio Program," in *Education on the Air: The Seventeenth Yearbook of the Institute for Education by Radio,* ed. O. Joe Olson (Columbus: Ohio State University, 1947), 463–74; Paul S. Lazarsfeld, "An Episode in the History of Social Research: A Memoir," in *The Intellectual Migration: Europe and America, 1930–1960,* ed. Donald Fleming and Bernard Bailyn (Cambridge, Mass.: Harvard University Press, 1969), 270–337; Mark R. Levy, "The Lazarsfeld-Stanton Program Analyzer: An Historical Note," *Journal of Communication* 32.4 (1982): 30–38; Wilson, "Effectiveness of Documentary Broadcasts."

85. John Gray Peatman, *The Patterning of Listener Attitudes toward Radio Broadcasts: Methods and Results* (Stanford, Calif: American Association for Applied Psychology/Stanford University Press, 1945), 8; Hallonquist and Peatman, "Diagnosing Your Radio Program."

86. Tore Hallonquist and Edward A. Suchman, "Listening to the Listener," in *Radio Research 1942–1943,* ed. Paul F. Lazarsfeld and Frank N. Stanton (1944; reprint, New York: Arno, 1979), 265, 292; Oscar Katz, "The Program Analyzer as a Tool," in *Education on the Air: Fifteenth Yearbook of the Institute for Education by Radio,* ed. Josephine H. MacLatchy (Columbus: Ohio State University, 1944), 266.

87. Tore Hallonquist to Edward R. Murrow, May 23, 1947, Shayon Papers (box 3, folder 12), 1–5; Wilson, "Effectiveness of Documentary Broadcasts."

88. Shayon, *Odyssey in Prime Time,* 108.

89. Robert Lewis Shayon interviewed by Erik Barnouw, January 26, 1967, transcript in Shayon Papers (box 77, folder 2).

90. Wilson, "Effectiveness of Documentary Broadcasts," 23.

91. Leonard C. Kercher, "Social Problems on the Air: An Audience Study," *Public Opinion Quarterly* 11.3 (1947): 408.

92. "Tense Anticipation Awaits 'New Light on Lincoln?' CBS Documentary at Unsealing of President's Papers," CBS news release, July 22, 1947, CBS News Archives; "CBS Gives Up 20G to Air Documentary," *Variety*, July 19, 1947, 6; "Close-Up of War Areas Today Will Stir G.I. Memories on Columbia's Global Documentary, 'We Went Back,'" CBS news release, July 25, 1947, CBS News Archives.

93. Robert Heller, "Actuality Broadcasts—'Slice-of-Life' Technique," *Variety*, January 8, 1947, 112. See also "Cross-the-Board Showcasing for CBS Tape Recorder," *Variety*, September 24, 1947, 21, 34.

94. "We Went Back," *Billboard*, August 23, 1947, 12, 16; "We Went Back," *Variety*, August 20, 1947, 31. CBS would drop its recording ban entirely the following year; see "Lifting of CBS Ban on Disked Shows Comes Suddenly; NBC Sole Holdout," *Variety*, November 17, 1948, 25.

95. Arnold Perl, *Fear Begins at Forty*, final broadcast script, CBS News Archives, 26; Sol Panitz, *Among Ourselves*, script reprinted in *Senior Scholastic*, February 23, 1948, 19. The Documentary Unit also produced a program the following spring on the plight of the American Indian in cooperation with the CBS Minneapolis affiliate WCCO. See "WCCO to Do a Doc on Indians for CBS," *Billboard*, February 14, 1948, 8; "Arrows in the Dust," *Variety*, May 26, 1948, 32; *Arrows in the Dust*, script in Sig Mickelson Papers (box 4Zd472, folder "WCCO Radio Scripts—1948"), Dolph Briscoe Center for American History, University of Texas at Austin.

96. "Ed Murrow Giving Up CBS Exec Berth; Returning to Air in Triple Pay Deal," *Variety*, July 16, 1947, 21; "Deny Murrow Friction," *Variety*, July 23, 1947, 27.

97. Charles Wertenbaker, "Profiles: The World on His Back," *New Yorker*, December 26, 1953, 36. For one account of William Shirer's departure from CBS, see Persico, *Edward R. Murrow*, 251–56.

98. Taylor telephone interview with author, June 30, 2008.

99. Ibid.; Taylor interview with Biagi, September 11, 1991, 51–52.

100. Robert P. Heller, "Videomentaries," *Variety*, July 28, 1948, 41. See also "CBS Program Shifts Linked to 'Agency' vs. 'Network' Situation? Heller Upped to Web's No. 2 Spot," *Variety*, June 2, 1948, 23.

101. "Radio Too 'Documentary-Happy'? CBS in Quest of a New Formula," *Variety*, September 15, 1948, 25, 36. See also "Michel Taking Over CBS Documentary Unit; May Expand into TV," *Variety*, June 23, 1948, 22; Jack Gould, "CBS to Repeat 'Mind in the Shadow,'" *New York Times*, February 13, 1949, 11.

102. Kercher, "Social Problems on the Air," 408, 410.

103. Saul Carson, "Notes toward an Examination of the Radio Documentary," *Hollywood Quarterly* 4.1 (1949): 73–74; "Radio Production," 381.

104. Charles A. Siepmann and Sidney Reisberg, "'To Secure These Rights': Coverage of a Radio Documentary," *Public Opinion Quarterly* 12.4 (1948–49): 650.

105. Shayon, *Odyssey in Prime Time*, 144.

106. Norman Corwin interviewed by Douglas Bell, *Years of the Electric Ear* (Metuchen, N.J.: Directors Guild of America/Scarecrow, 1994), 87–89; Shayon, *Odyssey in Prime Time*, 144. See also Smith, *In All His Glory*, 235–36.

107. "FCC Blue Book Seen Dead Duck; NAB Takes 'Who's Afraid?' Stance," *Variety*, September 24, 1947, 20; "Inside Stuff—Radio," *Variety*, April 2, 1947, 36. See also Llewellyn White, *The American Radio* (Chicago: University of Chicago Press, 1947); "On Impractical Do-Gooders," *Billboard*, April 12, 1947, 6, 12.

108. Heller, "Reporting by Radio," 11; Levy, "Lazarsfeld-Stanton Program Analyzer," 30.

109. See, for example, Todd Gitlin, "Media Sociology: The Dominant Paradigm," *Theory and Society* 6.2 (1978): 205–53.

110. Smith, *In All His Glory*, 153.

111. Ira Skutch, ed., *Five Directors: The Golden Years of Radio* (Lanham, Md.: Scarecrow, 1998), 177.

112. See Gilbert, *Cycle of Outrage*.

113. See Horwitt, *Let Them Call Me Rebel*, 532–36.

114. Paul Boyer, *By the Bomb's Early Light: American Thought and Culture at the Dawn of the Atomic Age* (1985; reprint, Chapel Hill: University of North Carolina Press, 1994), 300.

115. Susan J. Douglas, *Listening In: Radio and the American Imagination* (New York: Times Books, 1999), 164. There were a handful of female network-radio correspondents in the 1930s and 1940s. See Bliss, *Now the News*, 97–101; Douglas Gomery, *A History of Broadcasting in the United States* (Malden, Mass.: Blackwell, 2008), 42–47.

116. "Radio Blue Book Makes Good," *Variety*, March 12, 1947, 41.

117. Wilson, "Effectiveness of Documentary Broadcasts," 29.

118. Saul Carson, "Lazarsfeld-Field Survey Okay, but Was Interpretation Angled?" *Variety*, October 16, 1946, 31, 38.

119. Columbia University Bureau of Applied Social Research, *The People Look at Radio* (Chapel Hill: University of North Carolina Press, 1946), 10, 12. See also White, *American Radio*, 118–21; Pickard, *Media Democracy Deferred*, 63–65.

120. Lenthall, *Radio's America*, 152–53.

121. Shayon, "Eagle's Brood," 402.

122. Robert Lewis Shayon, "Talk before the Progressive Citizens of America Radio Division," May 20, 1947, Shayon Papers (box 18, folder 84). Shayon gave a similar talk at a PCA conference the following October; see chapter 6.

123. See A. M. Sperber, *Murrow: His Life and Times* (1986; reprint, New York: Fordham University Press, 1998), 277–78.

124. "Play Ball!" *Variety*, April 21, 1948, 29.

125. Wertenbaker, "Profiles," 29.

126. Charles A. Siepmann, "Radio Starts to Grow Up," *The Nation,* December 27, 1947, 697.

Chapter 4: Home Is What You Make It

1. Jack Gould, "A Hopeful Outlook," *New York Times,* August 10, 1947, 7; David Bird, "Kenneth R. Dyke Is Dead at 81; Helped to Democratize Japanese," *New York Times,* January 18, 1980, B5.

2. "Home Is What You Make It," note in Louis J. Hazam Papers (box 1, folder 1), Library of American Broadcasting, University of Maryland, College Park (hereafter Hazam Papers).

3. "Heritage of Home," October 6, 1945, script in Hazam Papers (box 1, folder 1).

4. "The Forgotten Age," October 27, 1945; "Thanksgiving Preview," November 17, 1945; "Your World Neighbors," December 1, 1945; all scripts in Hazam Papers (box 1, folders 3, 6, 8).

5. "Living," note in Hazam Papers (box 1, folder 38).

6. Louis J. Hazam biography, accessed August 10, 2010, http://www.lib.umd.edu/LAB/COLLECTIONS/hazam.html.

7. Sidney Lohman, "Radio Row: One Thing and Another," *New York Times,* February 15, 1948, 11.

8. Among the presidential candidates who appeared in the *Living 1948* time slot were Robert Taft, Earl Warren, Henry Wallace, and the Socialist candidate Norman Thomas.

9. "Living—1948," *Variety,* March 3, 1948, 28.

10. Susan Ohmer, *George Gallup in Hollywood* (New York: Columbia University Press, 2006), 5.

11. "The Mental State of the Nation," March 7, 1948, script in Hazam Papers (box 1, folder 38).

12. "Of Rats and Men," March 14, 1948, script in Hazam Papers (box 1, folder 39).

13. Jack Gould, "Programs in Review," *New York Times,* March 14, 1948, 9.

14. "From a Gentleman in Mufti," April 4, 1948; "New Draft, New Army," August 29, 1948; recordings from Jerry Haendiges Productions (http://www.otrsite.com).

15. Joseph E. Persico, *Edward R. Murrow: An American Original* (1988; reprint, New York: Laurel, 1990), 246.

16. "As Europe Sees Us," April 11, 1948, recording from Haendiges Productions.

17. "Then and Now," May 9, 1948; "American Self-Portrait, 1948," July 4, 1948; both recordings from Haendiges Productions.

18. "Plight of Our Hospitals: A Drama Diagnosis," September 5, 1948, recording from Haendiges Productions.

19. "Danger—School Zone," September 19, 1948, recording from Haendiges Productions. See also Benjamin Fine, *Our Children Are Cheated: The Crisis in American Education* (New York: Henry Holt, 1947).

188 · NOTES TO CHAPTER 4

20. "Freedom Is a Home-Made Thing," May 30, 1948; "The Sun and You," August 8, 1948; "USA—Growing Pains," April 18, 1948; all recordings from Haendiges Productions.

21. "Silver Cords and Apron Strings," March 21, 1948, recording from Haendiges Productions.

22. "Home Broken Home," April 25, 1948, recording from Haendiges Productions.

23. "Wisdom in the Streets," July 18, 1948, recording from Haendiges Productions.

24. Cabell Phillips, "And the Polls: What Went Wrong?" *New York Times*, November 7, 1948, E4. See also Kenneth Campbell, "Election Prophets Ponder in Dismay," *New York Times*, November 4, 1948, 8.

25. "What Happened?" November 7, 1948, script in National Broadcasting Company Records (box 326, folder 11), Wisconsin Historical Society, Madison (hereafter NBC Records). In the program, Gallup did not mention what would come to be the most-cited reason for the incorrect predictions—the pollsters had ended their surveys more than two weeks before the election, thus failing to catch late swings in Truman's favor. See Joseph C. Goulden, *The Best Years: 1945–1950* (New York: Atheneum, 1976), 421.

26. "Drama Documents," *Newsweek*, September 20, 1948, 62, 64. See also "NBC Off Public Service Hook: Documentaries Splurge Readied," *Variety*, August 18, 1948, 23, 30.

27. Jack Gould, "Programs in Review," *New York Times*, October 5, 1947, 11; "Wanted: A Baby," *Variety*, December 17, 1947, 30. See also Dick Doan, "Long Pants for Documentaries: Networks Vying in Public Service," *Variety*, June 25, 1947, 25, 30.

28. "The Atom and You," *Variety*, September 22, 1948, 24. See also "Atom with a Cherry on Top," *Time*, October 4, 1948, 42. Elsie Dick, unique in her position as a woman in charge of a network's documentaries, was killed along with several other journalists in a 1949 plane crash in India; see "Careers of Dead in India Air Crash," *New York Times*, July 13, 1949, 3.

29. See "Saudek Maps '48 Documentaries; Range from Communism to Ulcers," *Variety*, February 4, 1948, 26, 63.

30. "Skater on Thin Ice," *Newsweek*, June 7, 1948, 55.

31. Robert Saudek, "Public Interest, or How to Interest the Public," *Education* 67.9 (May 1947): 532–33; "Radio Production," in *Education on the Air: The Seventeenth Yearbook of the Institute for Education by Radio*, ed. O. Joe Olson (Columbus: Ohio State University, 1947), 376. See also "It's 'Waltz Me Around Again, Willie' As Durr, Saudek, White Bally Radio," *Variety*, March 3, 1948, 24, 34. For a fuller discussion of *Hiroshima* and *Unhappy Birthday*, see chapter 2.

32. "Radio Production," 373–76.

33. "School Teacher—1947," *Variety*, February 19, 1947.

34. "Radio Production," 374.

35. This and subsequent quotations are from *Slums: A Report on the Slums in America in Two Parts*, May 20, 1947, script in Erik Barnouw Papers (box 12), Rare Book and Manuscript Library, Columbia University, New York (hereafter Barnouw Papers).

36. Jack Gould, "Programs in Review," *New York Times*, May 25, 1947, 9.

37. "Slums," *Variety*, May 28, 1947, 30.

38. George Rosen, "Juve Problem (the Nation's Canker) Unveiled by CBS in Notable Broadcast," *Variety*, March 12, 1947, 46.

39. This and subsequent quotations are from *1960?? Jiminy Cricket!*, September 8, 1947, recording from Digital Deli Online (http://www.digitaldeliftp.com).

40. "1960?? Jiminy Cricket!" *Variety*, September 10, 1947, 28.

41. "Saudek Maps '48 Documentaries," 26.

42. "Saudek Sets ABC V.D. Documentary," *Variety*, April 14, 1948, 24.

43. The following discussion is taken from Erik Barnouw, *Media Marathon: A Twentieth-Century Memoir* (Durham, N.C.: Duke University Press, 1996), 93–111.

44. Ibid., 98, 101.

45. Ibid., 102–3.

46. Ibid.

47. These and subsequent quotations are from *V.D.: The Conspiracy of Silence*, April 29, 1948, script in Barnouw Papers (box 11).

48. Barnouw, *Media Marathon*, 103–9.

49. Jack Gould, "Programs in Review," *New York Times*, May 9, 1948.

50. "V.D.," *Variety*, May 5, 1948, 38.

51. "Commie Series on ABC Agenda," *Variety*, May 12, 1948, 27; "The Talk of the Town: Notes and Comment," *New Yorker*, July 3, 1948, 15.

52. Morton Wishengrad to Erik Barnouw, March 24, 1945, Barnouw Papers (box 38).

53. Morton Wishengrad autobiographical statement, Morton Wishengrad Papers (box 1, folder 17), Ratner Center for the Study of Conservative Judaism, Jewish Theological Seminary, New York (hereafter Wishengrad Papers).

54. Bob Lauter, "Around the Dial: ABC Network Introduces Red-Baiting under Heading of 'Documentary,'" *Daily Worker*, June 1, 1948, 13.

55. Qtd. in Morton Wishengrad to Anton Leader, July 22, 1948, Wishengrad Papers (box 1, folder 58).

56. Morton Wishengrad to Robert Saudek, July 28, 1948, Wishengrad Papers (box 1, folder 17). See also Robert Saudek to Morton Wishengrad, July 6, 1948, Wishengrad Papers (box 1, folder 17). The two men resolved their differences after the broadcast, and Wishengrad received his additional compensation.

57. Wishengrad to Leader, July 22, 1948, Wishengrad Papers.

58. These and subsequent quotations are from *Communism—U.S. Brand*, August 8, 1948, encore broadcast recording from Radio Program Archive, University of Memphis, Tenn.

59. Jack Gould, "Communism—U.S. Brand," *New York Times,* August 8, 1948, 7; "'We Do Not Question,'" *Time,* August 9, 1948, 39; "Communism—ABC Brand," *Newsweek,* August 16, 1948, 53; "Communism—U.S. Brand," *Variety,* August 4, 1948, 24. The program would win a Peabody Award the following year.

60. Bob Lauter, "ABC's 'Documentary' on Communism—1," *Daily Worker,* August 5, 1948, 13; Bob Lauter, "ABC's 'Documentary' on Communism—2," *Daily Worker,* August 8, 1948, 12; Bob Lauter, "ABC's 'Documentary' on Communism—3," *Daily Worker,* August 9, 1948, 13. See also "The Busy Air," *Time,* September 6, 1948, 75.

61. See "Saul Carson Dead; U.N. News Reporter," *New York Times,* June 22, 1971, 38; Sydney Gruson, "Tarle and Vavilov Will Lead Russians Attending World Congress of Intellectuals," *New York Times,* August 18, 1948, 10. See also Alonzo L. Hamby, *Beyond the New Deal: Harry S. Truman and American Liberalism* (New York: Columbia University Press, 1973), 195–97; Richard J. Walton, *Henry Wallace, Harry Truman, and the Cold War* (New York: Viking, 1976), 249–73; Mary Sperling McAuliffe, *Crisis on the Left: Cold War Politics and American Liberals, 1947–1954* (Amherst: University of Massachusetts Press, 1978), 33–47.

62. Saul Carson, "Half-Cocked Triggers," *New Republic,* July 12, 1948, 27.

63. Saul Carson, "The Network Sees Red," *New Republic,* August 16, 1948, 27–28.

64. Qtd. in Howard Blue, *Words at War: World War II Era Radio Drama and the Postwar Broadcasting Industry Blacklist* (Lanham, Md.: Scarecrow, 2002), 17.

65. Norman Rosten to Erik Barnouw, ca. 1945, Barnouw Papers (box 38). Rosten was similarly quoted in the FCC Blue Book, saying that commercial sponsorship stunted radio: "Shall the singing commercial and the Lone Ranger inherit the earth?" See *Public Service Responsibility of Broadcast Licensees* (Washington, D.C.: Federal Communications Commission, 1946), 17.

66. Norman Rosten to Morton Wishengrad, August 9, 1948, Wishengrad Papers (box 2, folder 2).

67. Morton Wishengrad to editor of the *New Republic,* August 14, 1948 [marked "Not Sent"], Wishengrad Papers (box 1, folder 12); Morton Wishengrad to Robert Saudek, August 15, 1948, Wishengrad Papers (box 1, folder 17).

68. Morton Wishengrad to Norman Rosten, August 14, 1948, Wishengrad Papers (box 2, folder 2).

69. Norman Rosten to Morton Wishengrad, two letters ca. August 1948; Morton Wishengrad to Norman Rosten, August 24, 1948; all in Wishengrad Papers (box 2, folder 2).

70. Morton Wishengrad to Norman Rosten, August 31, 1948; Norman Rosten to Morton Wishengrad, ca. September 1948; both in Wishengrad Papers (box 2, folder 2).

71. Morton Wishengrad to Norman Rosten, September 5, 1948 [two letters, the first marked "Not Sent"]; both in Wishengrad Papers (box 2, folder 2).

72. Norman Rosten to Morton Wishengrad, ca. September 1948, Wishengrad Papers (box 2, folder 2).

73. Robert Lewis Shayon, *Odyssey in Prime Time* (Philadelphia: Waymark, 2001), 110–12.

74. "CBS Comes Up With New 'On Tap' Dramatizations on Historical Events," *Variety*, April 9, 1947, 24.

75. Goodman Ace, "Top of My Head: I Was There," *Saturday Review*, September 20, 1969, 6; John Crosby, "Radio in Review: A New Slant on History," *New York Herald Tribune*, April 22, 1947, 20.

76. Robert Metz, *CBS: Reflections in a Bloodshot Eye* (Chicago: Playboy, 1975), 131–35.

77. This sample opening is taken from "The Citizens of Paris Before the Bastille," July 14, 1947, recording from Digital Deli Online. The same language was used in other episodes. For more on CBS executives' concerns about unduly alarming listeners, see Shayon, *Odyssey in Prime Time*, 117–18.

78. These and subsequent quotations are from "The Assassination of President Lincoln at Ford's Theatre," originally broadcast July 7, 1947, recording from Digital Deli Online. The Digital Deli website notes that there has been considerable confusion over the dates of certain recordings of *CBS Is There* (including that of the Lincoln show), given that numerous episodes aired more than once during the series' original run and that a few were rebroadcast later on the Armed Forces Radio Service. In addition, the beginnings of some episodes (including the Lincoln show) were altered to change the words "CBS Is There" to "You Are There." The available recordings otherwise appear to be an accurate representation of the content of the original series. The dates given here for each episode are taken from the information provided by Digital Deli; see http://www.digitaldeliftp.com/DigitalDeliToo/dd2jb-CBS-Is-There.html.

79. "'CBS Is There' Comes Back through Public Demand, Teeing Off This Month," *Variety*, October 8, 1947, 23; "*CBS Is There*: 'Radio by Popular Demand,'" CBS promotional brochure, March 1948, Robert Lewis Shayon Papers (box 9, folder 49), Howard Gotlieb Archival Research Center, Boston University (hereafter Shayon Papers); R. W. Stewart, "Programs in Review," *New York Times*, July 13, 1947, 55; Jerry Franken, "CBS Is There," *Billboard*, July 19, 1947, 15; "CBS Is There," *Variety*, July 9, 1947, 72.

80. René MacColl, "Eardrums along the Mohawk," *Atlantic Monthly*, May 1948, 92–94; Bernard DeVoto, "The Easy Chair," *Harper's* 195.1168 (September 1947): 250.

81. Shayon, *Odyssey in Prime Time*, 114–17; Ira Skutch, ed., *Five Directors: The Golden Years of Radio* (Lanham, Md.: Scarecrow, 1998), 177.

82. Ben Gross, "Looking and Listening: Priscilla No Pilgrim; Radio Man Tells Why," *New York Daily News*, March 28, 1948, sec. 2, 18.

83. Franken, "CBS Is There," 15.

84. Shayon, *Odyssey in Prime Time*, 118–19; Robert Lewis Shayon to Lester Haverling, May 31, 1974, Shayon Papers (box 115, folder 2).

85. Susan J. Douglas, *Listening In: Radio and the American Imagination* (New York: Times Books, 1999), 197–98; Bliss, *Now the News*, 107; "CBS Is There," *Variety*, October 29, 1947, 28.

86. See Jack Gould, "Television in Review," *New York Times*, October 9, 1949, 11; Robert E. Tomasson, "John Daly, Newsman, Dies at 77; Host of TV's 'What's My Line?'" *New York Times*, February 26, 1991, D23.

87. Unidentified review of *CBS Is There*, Shayon Papers (box 115, scrapbook). Douglas Edwards assumed John Daly's central role after Daly left the series in 1949.

88. "The Battle of Plassey," originally broadcast April 18, 1948. Quotations are taken from an undated rebroadcast of the program on the Armed Forces Radio Service. Recording from Digital Deli Online.

89. Ibid.

90. Gross, "Looking and Listening," sec. 2, 18; "A CBS Entry for the Variety 1948–49 Showmanagement Awards," ca. 1949, Shayon Papers (box 9, folder 49); Shayon, *Odyssey in Prime Time*, 119.

91. "Philadelphia, July 4, 1776," originally broadcast March 21, 1948, and rebroadcast July 4, 1948; "Lee and Grant at Appomattox," November 7, 1948. Recordings from Digital Deli Online.

92. "The Burr-Hamilton Duel," January 11, 1948, script in Shayon Papers (box 24, folder 112); "The Defense of the Alamo," August 11, 1947, recording from Digital Deli Online. For more on the debate over the Marshall Plan, see Eric F. Goldman, *The Crucial Decade—And After: America, 1945-1960* (New York: Vintage, 1960), 66–81.

93. "The Dreyfus Case," February 8, 1948; "The Betrayal of Toussaint L'ouverture," originally broadcast May 30, 1948, and rebroadcast January 23, 1949; "The Surrender of Sitting Bull," originally broadcast May 2, 1948, and rebroadcast January 2, 1949. Recordings from Digital Deli Online.

94. "CBS Makes History," unidentified and undated clipping, Shayon Papers (box 115, scrapbook); "Columbus Discovers America," October 10, 1948, recording from Digital Deli Online.

95. "The Witchcraft Trials at Salem," July 28, 1947; "The Trial of Anne Hutchinson," September 26, 1948, recordings from Digital Deli Online.

96. "The Death of Socrates," March 14, 1948, recording from Digital Deli Online. John Crosby, "Radio in Review: C.B.S. at Runnymede," *New York Herald Tribune*, April 19, 1948, 16.

97. "*CBS Is There*: 'Radio by Popular Demand.'"

98. Sidney Lohman, "One Thing and Another," *New York Times*, May 2, 1948, 89.

99. Shayon, *Odyssey in Prime Time*, 117, 123–24; Gross, "Looking and Listening," sec. 2, 18.

100. "NBC Off Public Service Hook," 23; Gould, "Communism—U.S. Brand," 7.

101. Douglas, *Listening In*, 33.

102. Shayon, *Odyssey in Prime Time*, 117–18. Shayon was not a fan of the 1950s television version of *You Are There*, calling it "a slick, picture-post card job" (119).

103. John Beaufort, "Air Historian Who Gets Around," *Christian Science Monitor*, March 8, 1948, n.p., Shayon Papers (box 115, scrapbook); "Time Machine," *Time*, July 28, 1947, 54.

104. Michele Hilmes, *Only Connect: A Cultural History of Broadcasting in the United States* (Belmont, Calif.: Wadsworth, 2002), 133. See also Bruce Lenthall, *Radio's America: The Great Depression and the Rise of Modern Culture* (Chicago: University of Chicago Press, 2007), 152–53.

105. "Mutual in Middle of Censor Row on 'Rights' Show," *Variety*, February 25, 1948, 23, 34.

106. "The News of Radio: MBS Abandons Its Plan for Dramatization of Civil Rights Report Series," *New York Times*, February 23, 1948, 36; "The Deep South Ain't Skeered of Civil Rights Issue," *Variety*, March 3, 1948, 22, 32; Charles A. Siepmann and Sidney Reisberg, "'To Secure These Rights': Coverage of a Radio Documentary," *Public Opinion Quarterly* 12.4 (1948–49): 649–58.

107. "Television: Many Happy Returns?," November 14, 1948, script in NBC Records (box 326, folder 11). This particular part of the script was written in verse, echoing radio works of Corwin and others.

108. Hubbell Robinson Jr., "The Last Stand?," *Variety*, July 28, 1948, 41.

109. These and subsequent quotes are from "Here You Are," recording in Shayon Papers (box 92). See also "Actors Spoof 'You Are There' Headman Shayon with Surprise Production of 'There You Are,'" CBS news release, January 19, 1949, Shayon Papers (box 115, folder 2).

Chapter 5: The Quick and the Dead

1. Davidson Taylor, "Is the Documentary Dead?" *Variety*, July 27, 1949, 42; "Documentarians in Fadeout: Old Standbys Go Commercial," *Variety*, December 8, 1948, 25, 32.

2. "Citizen of the World," July 10, 1949, recording from Digital Deli Online (http://www.digitaldeliftp.com). See also R. LeRoy Bannerman, *On a Note of Triumph: Norman Corwin and the Golden Years of Radio* (New York: Carol Publishing Group, 1986), 200–207; "New Corwin-CBS Pact in Doubt," *Variety*, January 12, 1949, 20; "Corwin Nixing Offers to Stay Freelance," *Variety*, February 2, 1949, 22; "Corwin to U.N. Radio Division Global Projects," *Variety*, March 9, 1949, 27.

3. Robert Lewis Shayon, *Odyssey in Prime Time* (Philadelphia: Waymark, 2001), 124–25, 141–46; "CBS Streamlining Axes Shayon, Six Asst. Directors; Chester Heads News," *Variety*, July 6, 1949, 28, 36; "Shayon, Roland Among 175 Out in 1 1/2 Million CBS Slash," *Billboard*, July 9, 1949, 6.

4. "Chorus at Nets: 'Cut, Cut, Cut,'" *Variety*, April 20, 1949, 21; "CBS Saving $111,000 in Its Lopping Off of 'You Are There' Series," *Variety*, July 13, 1949, 27. In the fall of 1949, CBS did bring back the radio version of *You Are There* under a new production team organized through the CBS Documentary Unit for a handful of episodes airing only once a month. The series was canceled for good the following summer in 1950. The television version debuted in 1953 and was hosted by Walter Cronkite.

5. "Lifting of CBS Ban on Disked Shows Comes Suddenly; NBC Sole Holdout,"

Variety, November 17, 1948, 25; "NBC Reported Ready to Yield on Disk Ban to Ward Off New Raids," *Variety,* January 19, 1949, 21; "Out of Bondage Via Tape," *Variety,* March 2, 1949, 26.

6. William F. Brooks, "TV News Benefited by Mistakes in Radio," *Variety,* July 27, 1949, 96.

7. Promotional materials for *Voices and Events,* ca. 1950, National Broadcasting Company Records (box 313, folder 41), Wisconsin Historical Society, Madison (hereafter NBC Records).

8. Mary Sperling McAuliffe, *Crisis on the Left: Cold War Politics and American Liberals, 1947–1954* (Amherst: University of Massachusetts Press, 1978), 8.

9. Ibid., 33–47; Michael C. Keith and Mary Ann Watson, eds., *Norman Corwin's One World Flight: The Lost Journal of Radio's Greatest Writer* (New York: Continuum, 2009), 192–94.

10. Richard Gid Powers, *Not without Honor: The History of American Anticommunism* (New York: Free Press, 1995), 203.

11. Arthur Schlesinger Jr., *The Vital Center: The Politics of Freedom* (1949; reprint, Boston: Houghton Mifflin, 1962), 38–40, 104, 125–26, 130, 136.

12. Norman Corwin to Edelaine Harburg, November 19, 1951, in *Norman Corwin's Letters,* ed. A. J. Langguth (New York: Barricade, 1994), 137.

13. "Subversive Persons in U.N. Jobs, M'Carran Charges, Citing Corwin," *New York Times,* August 9, 1949, 12; "'Mad Dog,' Corwin Says," *New York Times,* August 9, 1949, 12; "M'Carran Misquoted," *New York Times,* August 10, 1949, 16. See also David Everitt, *A Shadow of Red: Communism and the Blacklist in Radio and Television* (Chicago: Ivan R. Dee, 2007), 142–57.

14. Elmore McKee to Sterling Fisher and T. R. Carskadon, ca. 1949, NBC Records (box 326, folder 7). See also Schlesinger, *Vital Center,* 10, 249–51; Sidney Hook, "Academic Integrity and Academic Freedom: How to Deal with the Fellow-Travelling Professor," *Commentary* 8 (October 1949): 329–39.

15. "Buffalo Minister Succeeds Reiland," *New York Times,* July 17, 1936, 15.

16. "St. George's Parish Marks 125th Year," *New York Times,* December 14, 1936, 26; "Nazis Denounced at Many Services," *New York Times,* November 28, 1938, 12; "Ban on Singer 'Pagan,' Clergymen Declare," *New York Times,* March 5, 1939, 29.

17. "U.S. Policy Change for Peace Urged," *New York Times,* November 12, 1941, 18; "U.S. Camps Urged to Train Pacifists," *New York Times,* November 13, 1941, 13; "Would Ban 'Jap,'" *New York Times,* February 10, 1942, 18; "Obliteration Raids on German Cities Protested in U.S.," *New York Times,* March 6, 1944, 1.

18. "Dr. M'Kee Will Quit St. George's Pulpit," *New York Times,* June 28, 1946, 44; "The Word from Bat Cave," *Newsweek,* December 18, 1950, 54.

19. Elmore M. McKee, *The People Act* (New York: Harper, 1955), 1–10.

20. Schlesinger, *Vital Center,* 156; "Warns on Mob Frenzy," *New York Times,* February 14, 1938, 8. See also McAuliffe, *Crisis on the Left,* 63–74.

21. "Supplementary Memorandum: Proposed NBC Radio Series," January 9, 1950;

Wade Arnold to Charles C. Barry, October 10, 1950; both in NBC Records (box 326, folders 5 and 7). The Arnold to Barry memo indicates that the National Social Welfare Agency served as another "financial angel" by underwriting a series of five *Living 1950* programs aired in connection with a White House Conference on Children.

22. Sterling W. Fisher to Leslie Harris, April 7, 1950; Thomas C. McCray to Sterling W. Fisher, May 25, 1950; both in NBC Records (box 326, folder 7).

23. Ed King audition report, February 9, 1950, NBC Records (box 326, folder 7).

24. Wade Arnold to Elmore McKee, July 11, 1950, NBC Records (box 326, folder 7). See also Elmore McKee to Jack Turner, "Evaluation of Criticisms of 'The People Act,'" February 19, 1952, Robert Saudek/*Omnibus* Collection (History Series 1), Wesleyan Cinema Archives, Wesleyan University, Middletown, Conn. (hereafter Saudek/*Omnibus* Collection).

25. "The Twentieth Century Fund: Proposal for NBC Radio Series 'Living Democracy,'" May 15, 1950, NBC Records (box 326, folder 7).

26. This and subsequent quotations are from "Miracle on the Mount," December 9, 1950, script in NBC Records (call no. M96–123, folder 1). See also McKee, *People Act,* 13–28.

27. "Partners in Velvet," December 16, 1950, script in NBC Records (call no. M96–123, folder 2); "Cloth of Many Colors," February 3, 1951, recording from Original Old Radio, (http://www.originaloldradio.com). See also Mary Blackford Ford, "The People Act," *Social Action* 17.4 (April 15, 1951): 16–20.

28. "The Sylvania Story," January 6, 1951; "The City That Refused to Die," February 10, 1951; and "Red Clay and Teamwork," February 24, 1951; recordings from Original Old Radio. See also Ford, "People Act," 11–16; McKee, *People Act,* 215–35.

29. "Crusade in Baltimore," January 27, 1951; and "The Women Did It," January 20, 1951; recordings from Original Old Radio. See also Ford, "People Act": 22–24, 28–29; McKee, *People Act,* 121–46.

30. "The Prairie Noel," December 23, 1950; "The Sun Shines Bright," December 30, 1950; and "As the Children Go," February 17, 1951; recordings from Original Old Radio. See also Ford, "People Act": 20–21, 24–26, 32–35; McKee, *People Act,* 99–118.

31. "Home Is What They Made It," January 13, 1951; and "Our Partner—the Public," March 3, 1951; recordings from Original Old Radio. See also Ford, "People Act": 20–22, 29–32.

32. "The People Act," *Variety,* December 13, 1950, 42.

33. McKee, *People Act,* 10, 117–18. McKee did report that good relations between Morganville and its French counterpart were eventually restored.

34. "B'casters Puzzled on Red Treatment; Absence of Govt. Directives Hurts," *Variety,* August 30, 1950, 19.

35. "Calling the Turn on Korea!" NBC news release, July 10, 1950, NBC Records (box 327, folder 2).

36. "Borders in the Balance," July 22, 1950, script in Louis J. Hazam Papers (box 2, folder 33), Library of American Broadcasting, University of Maryland, College Park.

37. NBC news release, ca. August 1950, NBC Records (box 327, folder 7).

38. "Malice in Wonderland," parts 2 and 4, October 7, 1950, and October 21, 1950, scripts in NBC Records (box 327, folder 8).

39. Qtd. in Ralph Engelman, *Friendlyvision: Fred Friendly and the Rise and Fall of Television Journalism* (New York: Columbia University Press, 2009), 46.

40. Fred W. Friendly, *Due to Circumstances beyond Our Control . . .* (New York: Random House, 1967), xvii.

41. Engelman, *Friendlyvision*, 56–60.

42. Shayon, *Odyssey in Prime Time*, 118.

43. Engelman, *Friendlyvision*, 58–59; Robert Trout to Catherine "Kit" Trout, July 8, 1948, and Robert Trout to Catherine "Kit" Trout, August 23, 1948, in Robert Trout Papers (box 4Ze360, folder "Fred W. Friendly 1948–1965"), Dolph Briscoe Center for American History, University of Texas at Austin (hereafter Trout Papers); "Lost and Found Sound: Studio Nine," *Robert Trout Remembered*, National Public Radio, July 9, 1999; accessed August 12, 2010, http://www.npr.org/news/specials/001113 .trout.html. Friendly and Murrow went on to produce additional *I Can Hear It Now* albums; one on the 1920s relied heavily upon dramatizations and re-creations. See A. M. Sperber, *Murrow: His Life and Times* (1986; reprint, New York: Fordham University Press, 1998), 322.

44. *I Can Hear It Now*, LP (New York: Columbia Masterworks, 1948).

45. "Runaway," *New Yorker*, January 15, 1949, 22–23; Engelman, *Friendlyvision*, 59–60.

46. Robert Trout, "Character: Add Fred Friendly," Trout Papers (box 4Ze360, folder "Fred W. Friendly 1948–65").

47. Fred Friendly to William Brooks and Frank McCall, April 4, 1949, NBC Records (box 287, folder 11).

48. Robert Trout to William Brooks, July 10, 1949, and Ann Gillis to William Brooks, July 23, 1949, both in NBC Records (box 287, folder 11). See also Engelman, *Friendlyvision*, 61–64.

49. Anthony Leviero, "Truman Orders Hydrogen Bomb Built for Security Pending an Atomic Pact; Congress Hails Step; Board Begins Job," *New York Times*, February 1, 1950, 1, 3.

50. "Hope to Preem NBC Special Atom Series," *Variety*, June 14, 1950, 37.

51. Paul Boyer, *By the Bomb's Early Light: American Thought and Culture at the Dawn of the Atomic Age* (New York: Pantheon, 1985), 349.

52. Kathy J. Corbalis, "'Atomic Bill' Laurence: He Reported the Birth of the A-Bomb," *Media History Digest* 5.3 (Summer 1985): 9. See also William L. Laurence, *Dawn over Zero: The Story of the Atomic Bomb* (New York: Knopf, 1946); William L. Laurence, *Men and Atoms* (New York: Simon and Schuster, 1959).

53. Laurence, *Dawn over Zero*, 224.

54. Robert Jay Lifton and Greg Mitchell, *Hiroshima in America: Fifty Years of Denial* (New York: Grosset/Putnam, 1995), 12–22, 51–52; Robert Karl Manoff, "Covering the

Bomb: The Nuclear Story and the News," *Working Papers* 10.3 (Summer 1983): 22. See also Spencer R. Weart, *Nuclear Fear: A History of Images* (Cambridge, Mass.: Harvard University Press, 1988), 98–103; Boyer, *By the Bomb's Early Light,* 340–49; Beverly Deepe Keever, *News Zero: The* New York Times *and the Bomb* (Monroe, Me.: Common Courage Press, 2004), 39–48; Amy Goodman and David Goodman, "Hiroshima Cover-up: How the War Department's Timesman Won a Pulitzer," CommonDreams .org, August 10, 2004; accessed August 12, 2010, http://www.commondreams.org/views04/0810-01.htm.

55. NBC press release, June 28, 1950, NBC Records (box 289, folder 10).

56. Laurence agreement with NBC, May 24, 1950, NBC Records (box 289, folder 10). Laurence also would receive a new RCA television and 2 percent of the retail sales of a record album of the documentary. See unknown author to Paul Lukas, August 2, 1950, NBC Records (box 289, folder 10); Joseph O. Meyers to J. Shute, January 11, 1951, NBC Records (box 313, folder 30). See also Howard Taubman, "Records: The Atom," *New York Times,* April 15, 1951, 104.

57. "The Sound of the Atom," *Newsweek,* July 10, 1950, 50–51.

58. "The Mushroom Cloud," *Time,* July 17, 1950, 66. As of 1947, Hope had the highest-rated series on radio, although by 1950 those ratings (along with those for network radio in general) were in decline; see chapter 2 of this volume. See also Erik Barnouw, *The Golden Web: A History of Broadcasting in the United States 1933–1953* (New York: Oxford University Press, 1968), 284–90.

59. Letter to Bob Hope, June 7, 1950, NBC Records (box 289, folder 10); draft of letter to Hope, n.d., Fred Friendly Papers (box 179), Rare Book and Manuscript Library, Columbia University, New York (hereafter Friendly Papers). Neither the draft nor the actual letter is signed. However, the draft is in Friendly's papers and is typed on the same kind of paper with the same sort of handwritten corrections as the script drafts of *The Quick and the Dead* that Friendly wrote. A copy of the actual letter is included in William Brooks's papers in the NBC records and contains virtually identical language to the draft in Friendly's papers.

60. Unknown author to A. H. Sexton and T. H. Phelan, June 23, 1950, NBC Records (box 289, folder 10); "Mushroom Cloud," 66.

61. See Frederic Jacobi, "About 'The Quick and the Dead,'" *New York Times,* July 16, 1950, 7; Engelman, *Friendlyvision,* 66; "Mushroom Cloud," 66; Unidentified author and recipient, November 22, 1950, NBC Records (box 289, folder 10).

62. W. S. Parsons to Fred Friendly, June 16, 1950, NBC Records (box 312, folder 37). The bomb was armed in flight to avoid accidentally detonating it in the event of a crash during takeoff. See also Boyer, *By the Bomb's Early Light,* 316–18.

63. "Baruch's Speech at Opening Session of U.N. Atomic Energy Commission," *New York Times,* June 15, 1946, 4. See also Larry G. Gerber, "The Baruch Plan and the Origins of the Cold War," *Diplomatic History* 6.1 (Winter 1982): 69–95.

64. *The Quick and the Dead* script drafts, Friendly Papers (box 179).

65. Bernard M. Baruch to Fred Friendly, July 4, 1950, and unidentified author to

Bernard Baruch, July 5, 1950; both in NBC Records (box 289, folder 10). See also Joseph O. Meyers to Alan Keyes, February 27, 1951, NBC Records (box 290, folder 35).

66. Fred Friendly to William Brooks and Joseph O. Meyers, August 1, 1950, and unidentified author to Paul Lukas, August 2, 1950; both in NBC Records (box 289, folder 10).

67. Jacobi, "About 'The Quick and the Dead,'" 7.

68. "Mushroom Cloud," 66; "'Quick' Was Slow to Find Time," *Variety*, July 12, 1950, 1.

69. Kenneth M. Hance to William Brooks, June 26, 1950, NBC Records (box 289, folder 10).

70. This and subsequent quotations are taken from recordings of *The Quick and the Dead* on electronic-transcription discs in the University of Illinois Archives, Urbana (discs 88–89, 95–98). The original series ran two hours and aired July 6, 13, 20, and 27, 1950; an edited ninety-minute version aired at a later date. The recordings used for this study were of the original unedited broadcast. Scripts of the series are in the Friendly Papers (box 179).

71. At least one listener was not amused. He wrote NBC "to object to the derisive implication that Stagg Field was the scene only of defeat for Chicago. The fact is that our teams won many more games there than they lost." See Nelson H. Norgren to Niles Trammel, ca. July 1950, NBC Records (box 289, folder 10).

72. According to the *New York Times*, this sound was achieved "by approaching a piece of pure uranium toward a real neutron counter in the Columbia University physics laboratory." See Jacobi, "About 'The Quick and the Dead,'" 7.

73. *The Quick and the Dead* script drafts, Friendly Papers (box 179). For a discussion of the Baruch and Gromyko plans, see Gerber, "Baruch Plan and the Origins of the Cold War." A magazine piece that appeared at the time of *The Quick and the Dead*'s premiere indicated that Friendly was still undecided as to what form the last segment would take, even though that segment was scheduled to air only three weeks from then. See "Sound of the Atom," 51.

74. See ad in *Variety*, July 12, 1950, 31, and mailer in NBC Records (box 289, folder 10).

75. "Inside Stuff—Radio," *Variety*, June 28, 1950, 41; Hank Shepard to William Brooks, June 14, 1950, and Jo Dine to Charles Denny, July 3, 1950; both in NBC Records (box 289, folder 10). Dine defended the "atom kits" by saying that they would "not detract from the dignity and importance of the program," adding that "any device which will attract a few more people to the program is justified."

76. Jack Gould, "Documentary Series over N.B.C. on Atomic and Hydrogen Bombs Gets Off to Fine Start," *New York Times*, July 7, 1950, 40; "The Quick and the Dead," *Variety*, July 12, 1950, 30; June Bundy, "The Quick and the Dead," *Billboard*, July 15, 1950, 8; Saul Carson, "Blow by Blow," *New Republic*, August 28, 1950, 22; William Brooks to Joseph H. McConnell and Charles R. Denny, July 25, 1950, NBC Records (box 289, folder 10).

77. Bob Lauter, "'The Quick and the Dead'—Plug for Atomic Weapons," *Daily Worker*, July 11, 1950, 10.

78. Peabody Awards, official citation, accessed August 12, 2010, http://www.peabody.uga.edu/winners/details.php?id=970.

79. Engelman, *Friendlyvision*, 68. Friendly's representative, John G. Gude, wrote William Brooks that matters had grown "intolerable": "NBC's refusal to schedule [*Who Said That?*] in anything but marginal time has done very serious damage to what we consider a valuable property." See J. G. Gude to William Brooks, August 19, 1949, NBC Records (box 287, folder 11).

80. Fred Friendly to William Brooks and Joseph Meyers, August 1, 1950, NBC Papers (box 312, folder 37).

81. "'Who Said That?' May Wind Up as Two-Network Show If CBS Gets AM Version," *Variety*, February 15, 1950, 23.

82. Sig Mickelson, *The Decade That Shaped Television News: CBS in the 1950s* (Westport, Conn.: Praeger, 1998), 43–45. See also Engelman, *Friendlyvision*, 68–73.

83. See Edward Bliss Jr., *Now the News: The Story of Broadcast Journalism* (New York: Columbia University Press, 1991), 233. Friendly's papers contain only a transcript of the last segment as it aired; there is no script draft. Some of the actualities used in the segment do sound scripted, such as those from the boy with cancer. Others include minor stumbles and hesitations and sound more extemporaneous, as though they were given spontaneously in response to a reporter's question.

84. "Edward R. Murrow, Candidates for Offices to Be Heard in 'A Report to the Nation—The 1950 Election,'" CBS news release, November 6, 1950, Friendly Papers (box 179).

85. Mickelson, *Decade That Shaped Television News*, 45; Engelman, *Friendlyvision*, 72–73.

86. Stanley Cloud and Lynne Olson, *The Murrow Boys: Pioneers on the Front Lines of Broadcast Journalism* (Boston: Houghton Mifflin, 1996), 288.

87. "Your World Neighbors," December 1, 1945, script in Hazam Papers (box 1, folder 8).

88. "NBC Reports on 1949," NBC news release, January 3, 1950, NBC Records (box 326, folder 5).

89. Elmore McKee, "Director's Report for 'The People Act,'" ca. November 1951, Saudek/*Omnibus* Collection (History Series 1).

90. "Saudek to Head Ford Foundation Radio-TV Workshop; Resigns ABC," *Variety*, August 15, 1951, 24; "Saudek to Ford," *Newsweek*, August 27, 1951, 80.

91. Kevin G. Barnhurst and John Nerone, *The Form of News: A History* (New York: Guilford, 2001), 208; McKee, *People Act*, 246.

92. McKee, *People Act*, 245–46.

93. Herbert J. Gans, *Deciding What's News: A Study of CBS Evening News, NBC Nightly News*, Newsweek, *and* Time (1979; reprint, New York: Vintage, 1980), 43–44.

94. See Boyer, *By the Bomb's Early Light*, 338–41, 349.
95. William L. Laurence, *The Hell Bomb* (New York: Knopf, 1951), 88.
96. Manoff, "Covering the Bomb," 22.
97. Boyer, *By the Bomb's Early Light*, 291–302.
98. See Weart, *Nuclear Fear*, 104–6.
99. McKee, *People Act*, 239.
100. "Edward R. Murrow with the News" script, February 5, 1948, *Edward R. Murrow Papers, 1927–1965, Microfilm Edition* (Sanford, N.C.: Microfilming Corporation of America, 1982), reel 23, sec. 343, p. 1042. See also David Lilienthal, *The Journals of David E. Lilienthal: The Atomic Energy Years, 1945–1950* (New York: Harper and Row, 1964), 283–95; "To Those of Little Thought," *Time*, February 16, 1948, 23–24; Alexander Kendrick, *Prime Time: The Life of Edward R. Murrow* (Boston: Little, Brown, 1969), 309.

Chapter 6: Hear It Now

1. Robert Lewis Shayon, *Odyssey in Prime Time: A Life in Twentieth-Century Media* (Philadelphia: Waymark, 2001), 125.
2. Ibid., 130–32; "Shayon Mulls ECA Paris Assignment," *Variety*, February 1, 1950, 27; Robert R. Mullen to Robert Lewis Shayon, March 15, 1950, Robert Lewis Shayon Papers (box 114, folder 6), Howard Gotlieb Archival Research Center, Boston University (hereafter Shayon Papers).
3. Robert Lewis Shayon to Robert Mullen, February 28, 1950, Shayon Papers (box 114, folder 7).
4. John Beaufort to Robert Lewis Shayon, March 12, 1950, Shayon Papers (box 114, folder 7).
5. Shayon, *Odyssey in Prime Time*, 132–34. See also "M'Carthy Insists Truman Oust Reds," *New York Times*, February 12, 1950, 5.
6. "As the Witch Hunt Spreads, Radio Wonders If It, Too, Will Get Burned," *Variety*, October 29, 1947, 25; Arthur Schlesinger Jr., *The Vital Center: The Politics of Freedom* (1949; reprint, Boston: Houghton Mifflin, 1962), 125–26.
7. Schlesinger, *Vital Center*, 204.
8. Report on Norman Corwin, December 1, 1947, in American Business Consultants *Counterattack* Research Files (box 27, folder 14-37), Robert F. Wagner Labor Archives, Tamiment Library, New York University; "Red Scare Numbing Video," *Variety*, August 17, 1949, 33; Sig Mickelson, *The Decade That Shaped Television News: CBS in the 1950s* (Westport, Conn.: Praeger, 1998), 62. See also Erik Barnouw, *The Golden Web: A History of Broadcasting in the United States, 1933–1953* (New York: Oxford University Press, 1968), 253–57.
9. Qtd. in "'Blacklist' Counter-Offensive," *Variety*, July 13, 1949, 34.
10. Mickelson, *Decade That Shaped Television News*, 62–64; David Everitt, *A Shadow of Red: Communism and the Blacklist in Radio and Television* (Chicago: Ivan R. Dee, 2007), 71–72.

11. American Business Consultants, *Red Channels: The Report of Communist Influence in Radio and Television* (New York: Counterattack, 1950), 1–2.

12. Ibid.; Everitt, *Shadow of Red*, 42–46; Barnouw, *Golden Web*, 265–66.

13. "Radio Followup," *Variety*, September 20, 1950, 36.

14. American Business Consultants, *Red Channels*, 2, 4.

15. "Red Bait Vetoed by AFRA in N.Y.," *Variety*, October 23, 1946, 89.

16. "Grauer vs. Grauer," *Variety*, April 23, 1947, 25; "Ben Grauer Day," *Variety*, October 25, 1950, 25.

17. Everitt, *Shadow of Red*, 74–76; "Gypsy Rose Lee Denies Legion's Commie Charge; ABC Refuses to Cancel," *Variety*, September 13, 1950, 1, 61; Jack Gould, "Network Rejects Protest by Legion," *New York Times*, September 13, 1950, 9; "Peabody Awards Given in Radio, TV," *New York Times*, April 27, 1951, 41.

18. The nine included Corwin, Robson, and Rosten in addition to Orson Welles, Alan Lomax, Arthur Laurents, Arthur Miller, Millard Lampell, and Langston Hughes. See Erik Barnouw, ed., *Radio Drama in Action* (New York: Rinehart, 1945).

19. Barnouw, *Golden Web*, 266; Erik Barnouw, *Media Marathon: A Twentieth-Century Memoir* (Durham, N.C.: Duke University Press, 1996), 174.

20. Barnouw, *Media Marathon*, 172–73; Barnouw, *Golden Web*, 267–83; Everitt, *Shadow of Red*, 116–41.

21. Everitt, *Shadow of Red*, 70.

22. Jack Gould, "The Loyalty Probe," *New York Times*, January 7, 1951, 91.

23. A. M. Sperber, *Murrow: His Life and Times* (1986; reprint, New York: Fordham University Press, 1998), 365. See also Everitt, *Shadow of Red*, 70–80.

24. See Everitt, *Shadow of Red*, for a critical account of the political leanings of some *Red Channels* listees (42–70).

25. Qtd. in R. LeRoy Bannerman, *On a Note of Triumph: Norman Corwin and the Golden Years of Radio* (New York: Carol Publishing Group, 1986), 220.

26. Norman Corwin to Patrick Murphy Malin, October 25, 1950, in *Norman Corwin's Letters*, ed. A. J. Langguth (New York: Barricade, 1994), 130–31; Norman Corwin interviewed by Douglas Bell, *Years of the Electric Ear* (Metuchen, N.J.: Directors Guild of America/Scarecrow, 1994), 135–36. See also Everitt, *Shadow of Red*, 193–94.

27. Joseph Julian, *This Was Radio: A Personal Memoir* (New York: Viking, 1975), 172–82.

28. Morton Wishengrad to Sol Levitas, August 31, 1950, Morton Wishengrad Papers (box 1, folder 64), Ratner Center for the Study of Conservative Judaism, Jewish Theological Seminary, New York (hereafter Wishengrad Papers). See also Jack Gould, "'Red Purge' for Radio, Television Seen in Wake of Jean Muir Ouster," *New York Times*, August 30, 1950, 33.

29. Morton Wishengrad to Francis E. Walter, August 11, 1955; Joe Julian to Morton Wishengrad, May 28, 1954; both in Wishengrad Papers (box 1, folder 8). See also Julian, *This Was Radio*, 182–202; Everitt, *Shadow of Red*, 186–88.

30. Mickelson, *Decade That Shaped Television News*, 65–66. See also "CBS Veepees for Heller, Underhill?" *Variety*, August 2, 1950, 25; Sperber, *Murrow*, 366–67.

31. "Heller Quits CBS for Cowan Post," *Variety*, December 27, 1950, 19; "Novins to Get Heller CBS Job," *Billboard*, January 20, 1951, 3; "Heller Inks for Pic Co. Script," *Billboard*, February 3, 1951, 4. See also Edward R. Murrow to Marsland Gander, February 3, 1954; "Introductions for Bob Heller," ca. February 1954; Robert Peace Heller to Edward R. Murrow, February 5, 1954; Robert Heller to Edward R. Murrow, January 1955; all in *Edward R. Murrow Papers, 1927–1965, Microfilm Edition* (Sanford, N.C.: Microfilming Corporation of America, 1982), reel 4, sec. 30, pp. 595–97.

32. "Commie Issue to Snare Spotlight at RWG Election," *Variety*, October 25, 1950, 25, 36; "From We the Undersigned to RWG Members West and Midwest," October 16, 1950, Shayon Papers (box 114, folder 7).

33. Robert Lewis Shayon, "To Whom It May Concern," originally written April 25, 1953, and updated June 1, 1953, Shayon Papers (box 114, folder 6); Shayon, *Odyssey in Prime Time*, 132–41; Robert Lewis Shayon, "The Counter-Attack," speech given to the Conference on Cultural Freedom and Civil Liberties sponsored by the Progressive Citizens of America, October 25, 1947, Shayon Papers (box 18, folder 83).

34. Shayon, "To Whom It May Concern."

35. Shayon, *Odyssey in Prime Time*, 141.

36. Vincent W. Hartnett to Robert Lewis Shayon, July 11, 1953, Shayon Papers (box 114, folder 6); "Value of Counterattack's 'Red Channels' Proved by One of Its Listees," *New Counterattack*, January 1, 1954, 3, Shayon Papers (box 114, folder 6).

37. Shayon, *Odyssey in Prime Time*, 143–45.

38. Ibid., 145.

39. Sperber, *Murrow*, 361–67; Mickelson, *Decade That Shaped Television News*, 65.

40. Edward R. Murrow, *In Search of Light: The Broadcasts of Edward R. Murrow, 1938–1961*, ed. Edward Bliss Jr. (New York: Knopf, 1967), 115, 120, 158.

41. Joseph E. Persico, *Edward R. Murrow: An American Original* (New York: Laurel, 1988), 335–36.

42. "Sunday with Murrow," script of unaired pilot program, March 27, 1949, *Edward R. Murrow Papers,* reel 44, sec. 472, pp. 541–42.

43. Alexander Kendrick, *Prime Time: The Life of Edward R. Murrow* (Boston: Little, Brown, 1969), 319.

44. "A Report to the Nation—The 1950 Elections," *Variety*, November 15, 1950, 28.

45. "CBS Mulls 'Report' TV Counterpart; Sponsor Nibbles Already Seen," *Variety*, December 13, 1950, 32; "CBS to 'Hear It Now' Weekly," *Variety*, December 6, 1950, 29. It also was reported that the CBS executive Hubbell Robinson originally came up with the idea for the series.

46. For more on Paley's relationship with Murrow, see Sally Bedell Smith, *In All His Glory: The Life of William S. Paley* (New York: Touchstone, 1991).

47. "CBS Mulls 'Report' TV Counterpart," 32; John Crosby, "News Magazine of the Air," *New York Herald Tribune*, December 27, 1950, 21; "Sustainers Axed for CBS

'Report,'" *Variety,* December 13, 1950, 34; "Never Did So Many Need to Know So Much," CBS news release, December 8, 1950, *Edward R. Murrow Papers,* reel 46, sec. 510, p. 125. There were reports that commercial sponsors were interested in the radio series, but actual sponsorship would wait until the television version debuted the following year.

48. "Never Did So Many Need to Know So Much"; Mickelson, *Decade That Shaped Television News,* 45.

49. These and subsequent quotations are from *Hear It Now,* December 15, 1950, recording in University of Illinois Archives, Urbana. Recordings of most of the original episodes are also available from Digital Deli Online (http://www.digitaldeliftp.com).

50. Edward R. Murrow to Colonel Pat Echols, December 12, 1950, *Edward R. Murrow Papers,* reel 37, sec. 350, p. 665.

51. See Edward R. Murrow to W. G. McBride, January 15, 1951, *Edward R. Murrow Papers,* reel 37, sec. 350, p. 641; "Some Facts about *Hear It Now,*" CBS Sales Presentation, April 5, 1951, *Edward R. Murrow Papers,* reel 37, sec. 350, pp. 618–24.

52. Leon Morse, "Hear It Now," *Billboard,* December 23, 1950, 8; Jack Gould, "News Summaries," *New York Times,* December 24, 1950, 51; John Crosby, "News Magazine of the Air," *New York Herald Tribune,* December 27, 1950, 21; Robert Lewis Shayon, "Scraps of Sound and History," *Saturday Review,* February 10, 1951, 30.

53. "Hear It Now," *Time,* December 25, 1950, 44; *Hear It Now,* January 5, 1951, recording in University of Illinois Archives.

54. Edward R. Murrow to Joseph B. McFadden, January 26, 1951, *Edward R. Murrow Papers,* reel 37, sec. 350, p. 640; "Hear It Now," *Newsweek,* December 25, 1950, 44.

55. See Everitt, *Shadow of Red,* 105–6; Loren Ghiglione, *CBS's Don Hollenbeck: An Honest Reporter in the Age of McCarthyism* (New York: Columbia University Press, 2008); Bill Leonard, *In the Storm of the Eye: A Lifetime at CBS* (New York: G. P. Putnam's Sons, 1987), 70.

56. John C. Twitty, "CBS' 'Hear It Now' Presents Voices," *News Workshop,* May 7, 1951, *Edward R. Murrow Papers,* reel 18, sec. 253, p. 981. See also "Hear It Now: Logs" and "Hear It Now: Index," Joseph and Shirley Wershba Papers (box 4La117), Dolph Briscoe Center for American History, University of Texas at Austin.

57. Kendrick, *Prime Time,* 329–30.

58. *Hear It Now,* January 19, 1951, recording in University of Illinois Archives.

59. *Hear It Now,* February 16, 1951; May 4, 1951; and June 1, 1951; recordings in University of Illinois Archives.

60. *Hear It Now,* March 23, 1951 and May 11, 1951, recordings in University of Illinois Archives.

61. *Hear It Now,* May 11, 1951; March 16, 1951; and March 23, 1951; recordings in University of Illinois Archives.

62. *Hear It Now,* February 23, 1951, and June 15, 1951, recordings in University of Illinois Archives.

63. *Hear It Now,* May 4, 1951; April 6, 1951; December 22, 1950; February 2, 1951; and January 26, 1951; recordings in University of Illinois Archives.

64. *Hear It Now,* December 22, 1950, and March 9, 1951, recordings in University of Illinois Archives.

65. *Hear It Now,* June 1, 1951, recording in University of Illinois Archives. See also Persico, *Edward R. Murrow,* 318.

66. *Hear It Now,* January 12, 1951, and March 23, 1951, recordings in University of Illinois Archives.

67. See Murrow, *In Search of Light,* 166–69; Persico, *Edward R. Murrow,* 289–93; Sperber, *Murrow,* 340–49.

68. *Hear It Now,* May 18, 1951; February 16, 1951; and May 11, 1951; recordings in University of Illinois Archives.

69. See, for example, Edward R. Murrow and Fred W. Friendly, eds., *See It Now* (New York: Simon and Schuster, 1955), 55–67.

70. *Hear It Now,* December 29, 1950; February 9, 1951; January 12, 1951; and May 18, 1951; recordings in University of Illinois Archives.

71. *Hear It Now,* June 15, 1951, and April 13, 1951, recordings in University of Illinois Archives.

72. *Hear It Now,* February 9, 1951, recording in University of Illinois Archives; "Some Facts about *Hear It Now.*"

73. *Hear It Now,* April 13, 1951, recording in University of Illinois Archives.

74. *Hear It Now,* April 20, 1951; April 27, 1951; May 11, 1951; May 18, 1951; and May 25, 1951; recordings in University of Illinois Archives.

75. *Hear It Now,* February 16, 1951, and April 6, 1951, recordings in University of Illinois Archives. See also Philip Seib, *Broadcasts from the Blitz: How Edward R. Murrow Helped Lead America into War* (Washington, D.C.: Potomac, 2006).

76. *Hear It Now,* April 20, 1951, recording in University of Illinois Archives.

77. *Hear It Now,* February 2, 1951; February 9, 1951; April 27, 1951; and May 4, 1951; recordings in University of Illinois Archives.

78. *Hear It Now,* May 25, 1951, and June 8, 1951, recordings in University of Illinois Archives.

79. *Hear It Now,* February 16, 1951, recording in University of Illinois Archives.

80. Qtd. in Sperber, *Murrow,* 447.

81. William T. Evjue to Niles Trammel, September 11, 1951, National Broadcasting Company Records (box 115, folder 61), Wisconsin Historical Society, Madison. See also Edwin R. Bayley, *Joe McCarthy and the Press* (Madison: University of Wisconsin Press, 1981), 125–75, 192–95.

82. *Hear It Now,* December 15, 1950, and December 22, 1950, recordings in University of Illinois Archives.

83. *Hear It Now,* February 16, 1951; April 13, 1951; and May 11, 1951; recordings in University of Illinois Archives.

84. *Hear It Now,* June 15, 1951, recording in University of Illinois Archives.

85. Hubbell Robinson to William Paley and Frank Stanton, June 20, 1951, and Fred Friendly to Hubbell Robinson, June 19, 1951; both in *Edward R. Murrow Papers,* reel 22, sec. 128, p. 318.

86. Fred Friendly, "Video to Mirror World as 'Mass Information Gazette,'" *Variety,* July 11, 1951, 45.

87. Wesley Price, "Murrow Sticks to the News," *Saturday Evening Post,* December 10, 1949, 152; Persico, *Edward R. Murrow,* 294; Sperber, *Murrow,* 350.

88. "CBS Plans Hour Video Newscast," *Billboard,* August 18, 1951, 1; Edward R. Murrow to Bill McBride, September 7, 1951, *Edward R. Murrow Papers,* reel 37, section 350, p. 593; Leo Mishkin, "Murrow Looks to Tomorrow on TV," *New York Morning Telegraph,* September 3, 1951, 1–2.

89. Murrow, *In Search of Light,* 363. For overviews of Murrow and Friendly's television career, see Kendrick, *Prime Time;* Sperber, *Murrow;* Persico, *Edward R. Murrow;* and Ralph Engelman, *Friendlyvision: Fred Friendly and the Rise and Fall of Television Journalism* (New York: Columbia University Press, 2009).

90. "Top of the News," *Newsweek,* December 3, 1951, 58.

91. Persico, *Edward R. Murrow,* 301; Barnouw, *Golden Web,* 284–90.

92. See Edward Bliss Jr., *Now the News: The Story of Broadcast Journalism* (New York: Columbia University Press, 1991), 233; Val Adams, "Programs in Review," *New York Times,* July 29, 1951, 81; Elmore M. McKee, *The People Act* (New York: Harper, 1955); "Peabody Awards Given in Radio, TV."

93. Thomas Doherty, *Cold War, Cool Medium: Television, McCarthyism, and American Culture* (New York: Columbia University Press, 2003), 164.

94. Sperber, *Murrow,* 402–3.

95. See ibid., 546–48; Michael D. Murray, *The Political Performers: CBS Broadcasts in the Public Interest* (Westport, Conn.: Praeger, 1994), 32–38; Kendrick, *Prime Time,* 418–20; "Top of the News," 58; Charles Wertenbaker, "Profiles: The World on His Back," *New Yorker,* December 26, 1953, 36.

96. Bliss, *Now the News,* 237.

97. Jack W. Mitchell, *Listener Supported: The Culture and History of Public Radio* (Westport, Conn.: Praeger, 2005), 62–63.

98. Lawrence W. Lichty, "Documentary Programs on U.S. Radio," in *Museum of Broadcast Communications Encyclopedia of Radio,* vol. 1, ed. Christopher H. Sterling (New York: Fitzroy Dearborn, 2004), 476.

99. Price, "Murrow Sticks to the News," 152.

100. Kendrick, *Prime Time,* 195.

101. Michael Schudson and Susan E. Tifft, "American Journalism in Historical Perspective," in *The Press,* ed. Geneva Overholser and Kathleen Hall Jamieson (New York: Oxford University Press, 2005), 27.

102. See A. William Bluem, *Documentary in American Television: Form, Function, Method* (New York: Hastings, 1965), 71.

103. Wertenbaker, "Profiles," 29.

104. Sperber, *Murrow*, 357.
105. Fred W. Friendly, *Due to Circumstances Beyond our Control . . .* (New York: Random House, 1967), 3.
106. Sperber, *Murrow*, 403.
107. Thomas Rosteck, *See It Now Confronts McCarthyism: Television Documentary and the Politics of Representation* (Tuscaloosa: University of Alabama Press, 1994), 184–87; *Hear It Now*, June 15, 1951, recording in University of Illinois Archives.
108. Murrow, *In Search of Light*, 247–48; Rosteck, *See It Now Confronts McCarthyism*, 187–89; *Hear It Now*, January 19, 1951, recording in University of Illinois Archives.

Chapter 7: Lose No Hope

1. A. William Bluem, *Documentary in American Television: Form, Function, Method* (New York: Hastings House, 1965), 71.
2. Robert P. Heller, "Reporting by Radio," *New York Times*, October 26, 1947, 11; Robert P. Heller, "The Dynamic Documentary," in *Radio and Television Writing*, rev. ed., ed. Max Wylie (New York: Rinehart, 1952), 383.
3. See, for example, James W. Carey, "The Problem of Journalism History" (1974), in *James Carey: A Critical Reader*, ed. Eve Stryker Munson and Catherine A. Warren (Minneapolis: University of Minnesota Press, 1997), 86–94; Kevin G. Barnhurst and John Nerone, *The Form of News: A History* (New York: Guilford, 2001); Michael Schudson, *The Power of News* (Cambridge, Mass: Harvard University Press, 1995); Clifford G. Christians, Theodore L. Glasser, Dennis McQuail, Kaarle Nordenstreng, and Robert A. White, *Normative Theories of the Media: Journalism in Democratic Societies* (Urbana: University of Illinois Press, 2009).
4. Bluem, *Documentary in American Television*, 71, 93–100. See also Joseph E. Persico, *Edward R. Murrow: An American Original* (New York: Laurel, 1988), 299–326; A. M. Sperber, *Murrow: His Life and Times* (1986; reprint, New York: Fordham University Press, 1998), 351–413.
5. "We Went Back," *Variety*, August 20, 1947, 31; Robert P. Heller, "Videomentaries," *Variety*, July 28, 1948, 41; Fred Friendly to Hubbell Robinson, June 19, 1951, *Edward R. Murrow Papers, 1927–1965, Microfilm Edition* (Sanford, N.C.: Microfilming Corporation of America, 1982), reel 22, sec. 128, p. 318.
6. Erik Barnouw, *Documentary: A History of the Non-Fiction Film*, 2d rev. ed. (New York: Oxford University Press, 1993); Patricia Aufderheide, *Documentary Film: A Very Short Introduction* (New York: Oxford University Press, 2007), 127.
7. David Hendy, "'Reality Radio': The Documentary," in *More Than a Music Box: Radio Cultures and Communities in a Multi-Media World*, ed. Andrew Crisell (New York: Berghahn, 2004), 167.
8. Schudson, *Power of News*, 56; Barnhurst and Nerone, *Form of News*, 19.
9. Robert Heller, "Actuality Broadcasts—'Slice-of-Life' Technique," *Variety*, January 8, 1947, 112.

10. Barnhurst and Nerone, *Form of News*, 113.

11. Susan J. Douglas, "A Search for Community: Alone, Yet Together, All Listening," *Newsday*, December 6, 1999, A37; Susan J. Douglas, *Listening In: Radio and the American Imagination* (New York: Times Books, 1999), 33. See also Michele Hilmes, *Radio Voices: American Broadcasting, 1922–1952* (Minneapolis: University of Minnesota Press, 1997), 11–33.

12. *Hear It Now*, January 19, 1951, recording in University of Illinois Archives, Urbana.

13. Aufderheide, *Documentary Film*, 44; Bruce Lenthall, *Radio's America: The Great Depression and the Rise of Modern Culture* (Chicago: University of Chicago Press, 2007), 178, 197. See also William Stott, *Documentary Expression and Thirties America* (1973; reprint, Chicago: University of Chicago Press, 1986), 80–91; Howard Blue, *Words at War: World War II Era Radio Drama and the Postwar Broadcasting Industry Blacklist* (Lanham, Md.: Scarecrow, 2002).

14. Qtd. in Ralph Engelman, *Friendlyvision: Fred Friendly and the Rise and Fall of Television Journalism* (New York: Columbia University Press, 2009), 46.

15. Sig Mickelson, *The Decade That Shaped Television News: CBS in the 1950s* (Westport, Conn.: Praeger, 1998), 45.

16. Norman Corwin speech, February 18, 1946, Norman Corwin Collection (box 40, folder 437), Special Collections, Thousand Oaks Library, Thousand Oaks, Calif.

17. Bluem, *Documentary in American Television*, 90; Thomas Rosteck, *See It Now Confronts McCarthyism: Television Documentary and the Politics of Representation* (Tuscaloosa: University of Alabama Press, 1994), 30–33, 184–87.

18. Hilmes, *Radio Voices*, 11–33.

19. Herbert J. Gans, *Deciding What's News: A Study of CBS Evening News, NBC Nightly News, Newsweek, and Time* (1979; reprint, New York: Vintage, 1980), 42–52. See also Susan Merrill Squier, "Communities of the Air: Introducing the Radio World," in *Communities of the Air: Radio Century, Radio Culture*, ed. Susan Merrill Squier (Durham, N.C.: Duke University Press, 2003), 1–35; Douglas, *Listening In*, 3–21; Michael Curtin, *Redeeming the Wasteland: Television Documentary and Cold War Politics* (New Brunswick, N.J.: Rutgers University Press, 1995), 216–45; Rosteck, *See It Now Confronts McCarthyism*, 196–98.

20. See Curtin, *Redeeming the Wasteland*; Chad Raphael, *Investigated Reporting: Muckrakers, Regulators, and the Struggle over Television Documentary* (Urbana: University of Illinois Press, 2005); Thomas A. Mascaro, "The Chilling Effects of Politics: CBS News and Documentaries during the Fin-Syn Debate in the Reagan Years," *American Journalism* 22.4 (2005): 69–97.

21. George Rosen, "Radio Must Reform—or Else," *Variety*, October 23, 1946, 1, 90; "Paley's Primer on Programming," *Variety*, October 23, 1946, 90; Jack Gould, "Programs in Review," *New York Times*, May 25, 1947, 9; Charles A. Siepmann and Sidney Reisberg, "'To Secure These Rights': Coverage of a Radio Documentary," *Public Opinion Quarterly* 12.4 (1948–49): 650.

22. Curtin, *Redeeming the Wasteland*, 248.

23. Alexander Kendrick, *Prime Time: The Life of Edward R. Murrow* (Boston: Little, Brown, 1969), 319; Mickelson, *Decade That Shaped Television News*, 43–45.

24. See Victor Pickard, "Media Democracy Deferred: The Postwar Settlement for U.S. Communications, 1945–1949," Ph.D. dissertation, University of Illinois at Urbana-Champaign, 2008; Victor Pickard, "Reopening the Postwar Settlement for U.S. Media: The Origins and Implications of the Social Contract between Media, the State, and the Polity," *Communication, Culture, and Critique* 3.2 (2010): 170–89; Elizabeth Fones-Wolf, *Waves of Opposition: Labor and the Struggle for Democratic Radio* (Urbana: University of Illinois Press, 2006), 150–61.

25. David Paul Nord, "A Plea for Journalism History," *Journalism History* 15.1 (1988): 10–11; James W. Carey, "'Putting the World at Peril': A Conversation with James W. Carey," in *James Carey: A Critical Reader*, ed. Eve Stryker Munson and Catherine A. Warren (Minneapolis: University of Minnesota Press, 1997), 109; James S. Ettema and Theodore L. Glasser, "The Irony in—and of—Journalism: A Case Study in the Moral Language of Democracy," *Journal of Communication* 44.2 (1994): 5.

26. James W. Carey, "The Press and the Public Discourse," *Center Magazine* 20.2 (March/April 1987): 14.

27. Bluem, *Documentary in American Television*, 90.

28. Richard Rorty, *Achieving Our Country: Leftist Thought in Twentieth-Century America* (Cambridge, Mass.: Harvard University Press, 1998), 106; Robert Lewis Shayon, "Talk before the Progressive Citizens of America Radio Division," May 20, 1947, Robert Lewis Shayon Papers (box 18, folder 84), Howard Gotlieb Archival Research Center, Boston University (hereafter Shayon Papers).

29. Thomas Doherty, *Cold War, Cool Medium: Television, McCarthyism, and American Culture* (New York: Columbia University Press, 2003), 174–77.

30. Carey, "'Putting the World at Peril,'" 109.

31. Robert Lewis Shayon, *Odyssey in Prime Time* (Philadelphia: Waymark, 2001), 147.

32. Nord, "Plea for Journalism History," 11; Michael Schudson, "Toward a Troubleshooting Manual for Journalism History," *Journalism and Mass Communication Quarterly* 74.3 (1997): 470–71.

33. Shayon, *Odyssey in Prime Time*, 98.

34. See, for example, Robert W. McChesney and John Nichols, *The Death and Life of American Journalism: The Media Revolution That Will Begin the World Again* (Philadelphia: Nation Books, 2010).

35. Ruth Ashton Taylor interview with Shirley Biagi for the Washington Press Club Foundation, September 11, 1991, 48; accessed August 6, 2010, http://wpcf.org/oralhistory/tay2.html.

36. *Operation Crossroads*, final broadcast script, May 28, 1946, Shayon Papers (box 6, folder 34).

37. Jay Allison, "Radio Storytelling Builds Community On-Air and Off," *Nieman*

Reports 55.3 (Fall 2001): 16–17. See also Matthew C. Ehrlich, "Poetry on the Margins: *Ghetto Life 101, Remorse,* and the New Radio Documentary," *Journalism: Theory, Practice, and Criticism* 4.4 (2003): 423–39; Douglas, *Listening In,* 22–39; Daniel Makagon and Mark Neumann, *Recording Culture: Audio Documentary and the Ethnographic Experience* (Los Angeles: Sage, 2009); John Biewen and Alexa Dilworth, eds., *Reality Radio: Telling True Stories in Sound* (Chapel Hill: University of North Carolina Press, 2010).

38. James S. Ettema and Theodore L. Glasser, *Custodians of Conscience: Investigative Journalism and Public Virtue* (New York: Columbia University Press, 1998), 200–201. See also John J. Pauly, "Journalism and the Sociology of Public Life," in *The Idea of Public Journalism,* ed. Theodore L. Glasser (New York: Guilford, 1999), 134–51.

39. See Gary Edgerton, "The Murrow Legend as Metaphor: The Creation, Appropriation, and Usefulness of Edward R. Murrow's Life Story," *Journal of American Studies* 15.1 (1992): 75–91.

40. Edward R. Murrow, *In Search of Light: The Broadcasts of Edward R. Murrow, 1938–1961,* ed. Edward Bliss Jr. (New York: Knopf, 1967), 357.

41. Jay Allison, "A Word from Jay Allison," *Transom Review* 1.14 (2001): 2; accessed August 20, 2010, http://www.transom.org/guests/review/200112.review.ncorwin.pdf. See also the accompanying comments from other radio practitioners responding to Corwin's work.

42. Norman Corwin, telephone interview with the author, February 12, 2006.

43. Ibid.; Norman Corwin, "One World Revisited," *Common Ground* 7.2 (1947): 17.

Index

Aaron, John, 142
ABC (American Broadcasting Company), 8, 9; creation of, 80; and recordings, 31, 80–82, 85, 100, 111; and *Red Channels*, 132–33; rivalry with other networks, 31, 79–80, 100, 160; and Saudek, Robert, 9, 79–92, 100, 125, 132, 160. *See also* ABC documentaries
ABC documentaries: *Communism—U.S. Brand*, 9, 87–93, 102; *Hiroshima* (radio program), 28–29, 80, 102; *1960??*; *Jiminy Cricket!*, 9, 82–84, 110, 157; *School Teacher—1947*, 80; *Slums*, 80–82, 111; *Unhappy Birthday*, 29, 58, 80; *V.D.: The Conspiracy of Silence*, 9, 84–87
Ace, Goodman, 93
Ace, Jane, 93
actuality documentary. *See* recordings
Acuff, Roy, 85
Adams, John, 97
Adler, Luther, 54–56, 131
Agee, James, 15
Agronsky, Martin, 148
Albert, Eddie, 65
Aldrich Family, The, 135
Alinsky, Saul, 9, 52–56, 68
Allison, Jay, 163–64
All Things Considered, 152
Alsop, Joseph, 148
Alsop, Stewart, 148
American Business Consultants, 130. See also *Counterattack*; *Red Channels*
American Civil Liberties Union, 134
American Federation of Radio Artists, 132
American in England, An, 1–2, 25
American Legion, 132, 136
American Radio, The (White), 67
Americans for Democratic Action (ADA), 106
Americans United for World Government, 29
America's Needs and Resources (Twentieth Century Fund), 83, 110
Among Ourselves, 65
Anderson, Benedict, 4
Anderson, Marian, 107
Armed Forces Radio Service, 18, 191n78
Arnold, Hap, 149
Arnold, Wade, 73, 113
Arrows in the Dust, 185n95
Ashton, Ruth: early career of, 57–58; and Heller, Robert, 59, 66; and Murrow, Edward R., 50, 58, 59, 65–66; and *Report Card*, 66, 70; and sexism, 57, 68; and *Sunny Side of the Atom*, 3, 9, 58–61, 162; and television news, 66
As Others See Us, 75
Assignment Home, 20
Atlantic, 95
Atom and You, The, 79
atomic energy and weapons in documentary: and *Atom and You*, 79; early radio treatments of, 28–29; and *Hear It Now*, 147–48; and *I Can Hear It Now*, 115, 126;

212 · INDEX

and *Operation Crossroads*, 50–51, 62, 162; and *Quick and the Dead*, 114–23, 126–28; and *Sunny Side of the Atom*, 58–61, 64, 68, 69, 162
Attlee, Clement, 33, 36, 168n34
audiotape. *See* recordings
Austin, Warren, 149

Bacher, Robert, 118
Back of the Yards Neighborhood Council (Chicago), 52–53, 157
Barber, Red, 139, 140, 141
Barnhurst, Kevin, 125, 156
Barnouw, Erik, 18, 22, 136, 156; and *Cavalcade of America*, 14–15; and *Red Channels*, 131, 133; and *V.D.: The Conspiracy of Silence*, 3, 9, 84–87
Baruch, Bernard, 118–19, 122, 126
Baruch Plan, 118–19, 122, 126
Battle of Midway, The, 18
Battle of San Pietro, The, 18
Battle of the Warsaw Ghetto, The, 18, 87, 91–92
BBC (British Broadcasting Corporation), 30, 47, 142
Beier, Carl, 59
Benes, Edvard, 37, 42
Benny, Jack, 67
Best Years of Our Lives, The, 22
"Big Annie." *See* Stanton, Frank: and Lazarsfeld-Stanton Program Analyzer
Billboard: and *Hear It Now*, 141, 151; and Heller, Robert, 136; and *One World Flight*, 39; and *Quick and the Dead*, 123; and *See It Now*, 151; and *We Went Back*, 64–65
blacklisting. *See Counterattack*; *Red Channels*
Blackwell, Lane, 50
Bland, Lee, 31–34, 176n55
Bliss, Edward, 20
Blue Book (*Public Service Responsibility of Broadcast Licensees*), 6, 69, 82; and Corwin, Norman, 41; and Durr, Clifford, 41, 131; impact of, 67–68, 159–60; origins of, 23, 47; and Paley, William, 9, 41, 47–48, 159–60; reaction to, 9, 41, 47–48; and Siepmann, Charles, 47, 70, 114, 160
Bluem, A. William, 3, 4, 155, 158
Booth, John Wilkes, 94–95
Born Yesterday (movie), 140

Bourke-White, Margaret, 15
Boyer, Paul, 116, 126
Brooks, William, 20, 105; and *Quick and the Dead*, 115–16, 117, 119, 123
Broun, Heywood, 173n16
Brown, Cecil, 21
Bunche, Ralph, 144
Burrows, Abe, 139, 140, 141

Calmer, Ned, 94, 96
Capital Times (Madison, Wisconsin), 148
Capra, Frank, 18, 26, 50
Carey, James, 5, 11, 160–61
Carney, Art, 75
Carson, Saul, 66–67, 91–92
Carson, Velma, 112
Cavalcade of America, The, 8, 14–15, 24
CBS (Columbia Broadcasting System): and budget cuts, 10, 104–5, 137; and liberalism, 21, 45, 67, 69–70; loyalty oath of, 133, 135, 138, 141; news operation of, 1, 16–17, 20; postwar public affairs series of, 20, 23, 48, 75; prewar programming of, 13–15, 16, 24–25; and radio commentators, 21; and radio research, 9, 61–64, 67–69; and recording ban, 30–31, 53, 64–65, 73, 105; and red-baiting, 6, 44–45, 130–31; and sustaining programming, 16, 32, 46–47, 160; and television, 21, 66, 102, 124, 137, 150–51; and World War II, 1–2, 16–20, 30, 46–47, 49. *See also* CBS documentary programs and series; CBS-NBC rivalry
CBS documentary programs and series, 8–10, 79, 159–60; *Among Ourselves*, 65; *Arrows in the Dust*, 185n95; baseball program, 70; *Cavalcade of America*, 14–15; *Citizen of the World*, 104–5, 131; *Eagles' Brood*, 9, 51–56, 68–70, 82, 131, 141; *Empty Noose*, 49, 131; *Experiment in Living*, 57, 70; *Fear Begins at Forty*, 65, 131; *Hear It Now*, 3, 10, 138–54; Lincoln documentaries, 64; *Long Life and a Merry One*, 57, 64; *March of Time* (radio series), 13–14; *Mind in the Shadow*, 66; *Nation's Nightmare*, 151; *On a Note of Triumph*, 8, 25, 26–27, 68; *One World Flight*, 8–9, 28–45, 48, 130, 152, 157; *Open Letter on Race Hatred*, 2, 18–19, 131, 174n26; *Operation Crossroads*, 29, 50–51; *People Act*, 125, 151; prostitution program, 152; *Report*

Card, 66; *See It Now*, 150–56; *Sunny Side of the Atom*, 58–61, 64, 68–70, 127; *We Went Back*, 64–65; *Who Killed Michael Farmer?*, 152; *You Are There* (radio series), 5, 9, 65–66, 93–103, 114, 157. See also *I Can Hear It Now*
CBS Documentary Unit. *See* CBS documentary programs and series
CBS Is There. See *You Are There* (radio series)
CBS–NBC rivalry, 46, 67, 102; and documentaries, 59, 71, 123–25, 160; and recording ban, 105; and sustaining programs, 16, 32; and television, 21
CBS Views the Press, 48, 139
Chiang Kai-shek, 27, 34
Chou En-lai, 27
Christian Century, 61
Christian Science Monitor, 101
Churchill, Winston, 14, 28, 149, 168n34
Citizen of the World, 104–5, 125, 131
Clark, E. Gurney, 84–85
Clark, Evans, 110, 111
Clark, Tom, 51
Clive, Robert, 97
Close, Upton, 21
Cobb, Lee J., 104, 131
cold war, 2, 157–60; escalation of, 73, 113; and *Hear It Now*, 147–50; and *Living*, 73, 75–76, 102, 113–14; origins of, 28; and *People Act*, 107–8, 113, 125–26; and television documentaries, 3; and *You Are There* (radio series), 98. See also *Communism—U.S. Brand*; *Counterattack*; Korean War; *One World Flight*; *Quick and the Dead, The*; *Red Channels*
Collingwood, Charles, 1, 21, 167–68n34
Columbia Records, 9, 115
Columbia University, 57, 63, 84–85, 87
Columbia Workshop, 16
commentators, 21, 22, 41
Commission on Freedom of the Press, 23
Committee on the Present Danger, 138
Common Council for American Unity, 28
Communism—U.S. Brand, 9, 100, 106, 131, 135; discussion of, 87–93, 102, 158
Copland, Aaron, 26
Corwin, Norman, 2, 46, 101–2, 154, 158, 161; and *American in England*, 1–2, 25; and Bill of Rights program, 25–26; and *Citizen of the World*, 104–5, 125, 131; and common man, 25–26, 38; and *Counterattack*, 2, 6, 44, 130–34; departure from CBS of, 6, 10, 67, 104; and Hope, Bob, 40–41, 117; and House Committee on Un-American Activities (HUAC), 44, 130; influences on, 26, 173n16; legacy of, 43, 163–64; McCarran charges against, 106–7; and Murrow, Edward R., 25, 42; and *On a Note of Triumph*, 8, 25, 26–27, 68; and *One World Flight*, 8–9, 28–45, 48, 130, 152, 157; and Paley, William, 10, 28, 31, 34, 67, 104; and *Passport for Adams*, 30; politics of, 26–29, 44, 69, 106; radio as viewed by, 6, 8, 23, 24, 41; and *Red Channels*, 2, 6, 44, 130–34; and Schlesinger, Arthur Jr., 106, 130; and sponsorship, 40–41, 134; and *They Fly through the Air with the Greatest of Ease*, 24–25; and *This is War!*, 25, 50; and *Twenty-six by Corwin*, 24; and World War II, 1–2, 24–27
Costello, Bill, 32, 44
Costello, Frank, 143
Cotten, Joseph, 9, 53–55
Counterattack: and Corwin, Norman, 44, 130; and *Red Channels*, 130–37; and Shayon, Robert Lewis, 129–30, 136–37
Cowan, Louis, 136
Cronkite, Walter, 193n4
Crosby, Bing, 31, 56, 105
Crosby, John, 93, 99, 139, 141
Cross Section-USA, 20
"Cultural Front," 5, 6
Custer, George Armstrong, 98

Daily Worker, 44, 123, 127; and *Communism—U.S. Brand*, 88–92
Daly, John, 115; and *You Are There* (radio program), 94–103
Davis, Benjamin J., 136
Davis, Elmer, 148
Dawn Over Zero (Laurence), 116
de Gaulle, Charles, 27, 115
delinquency. *See* juvenile delinquency and documentary
DeVoto, Bernard, 95
Dewey, Thomas, 78, 147
Diamond, David, 140, 142
Dichter, Ernest, 9, 61–63, 64, 67–68
Dick, Elsie, 79, 188n28

Dickinson, John, 97
DiMaggio, Joe, 143
Dine, Jo, 198n75
Disney, Walt, 83
Doctors at Home, 20
docudrama. *See* dramatized documentary
Doherty, Lorraine, 103, 105
Douglas, Susan J., 4, 96, 100, 157
Douglas, William O., 51
Downs, Bill, 138, 151
dramatized documentary: characteristics of, 5, 7–8, 53; criticisms of, 30, 64, 73; decline of, 153, 155–59; defenses of, 53, 65, 156
Dreyfus, Alfred, 98
Dubinsky, David, 88
DuPont, 8, 14–15, 16
Durr, Clifford, 8, 41, 48, 131
Dyke, Ken, 71, 72

Eagle's Brood, The, 3, 9, 57, 59, 93, 131; docudrama format of, 53, 156–57; and grassroots action, 51–53, 55, 82, 101, 107, 141; impact of, 55–56, 68–70; origins and production of, 50–55; and radio research, 62–64
Easy Aces, 93
Economic Cooperative Association (ECA), 129–30
editorializing, 70. *See also* objectivity
education and documentary: and *Living 1948*, 76–77; and *People Act*, 112; and *Report Card*, 66; and *School Teacher—1947*, 80
Edwards, Cliff, 83
Edwards, Douglas, 96, 97, 167n34
Einstein, Albert: and *Operation Crossroads*, 51, 162; and *Quick and the Dead*, 119–22; and *Sunny Side of the Atom*, 3, 58, 60, 69, 162
Einstein, Sergei, 33, 37
Eisenhower, Dwight D., 31, 75, 85, 147, 168n34
Eliot, George Fielding, 96
Ellery Queen, 49
Empty Noose, The, 49, 50, 53, 65, 66, 131
Enola Gay, 115, 118, 121, 126
Eternal Light, The, 87
Ettema, James, 161
Evans, Walker, 15
Evjue, William, 148–49

Experiment in Living, 57, 70

Fall of the City, The, 16
Fanfare for the Common Man (Copland), 26
FBI (Federal Bureau of Investigation), 130, 136
FCC. *See* Federal Communications Commission
Fear Begins at Forty, 65, 131
Feature Story, 57
Federal Communications Commission (FCC), 2, 8, 21, 22–23, 131; and Blue Book, 6, 9, 23, 41, 47–48, 68–69, 159–60; and editorializing, 70
Ferebee, Tom, 121
Fermi, Enrico, 120
Fine, Benjamin, 76–77
Fleming, James, 105
Fletcher, Lucille, 173n16
Flood, The, 15
Folkways Records, 4
Footsteps in the Sands of Time, 114
Ford, John, 18
Ford Foundation, 125, 160
Franco, Francisco, 18, 24
Frankel, Lou, 40
Freedom House, 28, 31
Friendly, Fred: early career of, 31, 114; and *Hear It Now*, 3, 10, 138–60; and *I Can Hear It Now*, 7, 9, 114–17, 124, 126, 151–52, 196n43; Murrow initial pairings with, 10, 114–15, 123–24, 139; and *Quick and the Dead*, 3, 10, 114–28; and recordings, 31, 117–19, 124, 138–54, 157; and *Report to the Nation—The 1950 Elections*, 139; and *See It Now*, 140, 150–61; and television, 10, 115, 124, 140, 150–61; and *Who Said That?*, 115, 123
Frontiers of Science, 20
Front Page, The (television series), 96
Fuchs, Klaus, 121
Fun and Fancy Free, 83

Gallup, George, 9, 73–79, 101
Gans, Herbert, 126, 159
Geer, Will, 77, 131
General Federation of Women's Clubs, 54
Ghandi, Mahatma, 14
Gitlin, Irving, 142, 151
Glasser, Theodore, 161

Glazer, Tom, 87, 131, 133, 135
Godfrey, Arthur, 103, 106
Gould, Jack, 47, 71; and Blue Book, 82, 160; and CBS loyalty oath, 133; and *Communism—U.S. Brand*, 90, 100, 102; and *Eagle's Brood*, 56; and *Hear It Now*, 141; and *Living 1948*, 74–75; and *One World Flight*, 40; and *Quick and the Dead*, 123; and *Slums*, 82
Grapes of Wrath, The (Steinbeck), 15
Grauer, Ben, 74–78, 109–10, 114, 131–32
Grierson, John, 15, 17
Groves, Leslie, 58, 118, 127
Gude, John G. "Jap," 114, 199n79

Hamilton, Alexander, 98
Hampden, Walter, 99
Harper's, 95
Hartnett, Vincent, 131, 137
Hayes, Helen, 119, 120
Hazam, Lou: and *Home Is What You Make It*, 9, 71–73, 112, 124; and *Living*, 113, 124, 161; and *Living 1948*, 9, 72–79; and *1960?? Jiminy Cricket!*, 83; and *People Act*, 109–13; and television documentary, 73
health care and documentary: and Fear Begins at Forty, 65; and *Living 1948*, 74, 76; and *Long Life and a Merry One*, 57, 64; and *Mind in the Shadow*, 66; and *People Act*, 109–10; and *V.D.: The Conspiracy of Silence*, 84–87
Hear It Now, 3, 10, 138–60
Heatter, Gabriel, 21
Heflin, Van, 57
Heirens, William, 51
Heller, Robert, 2, 50, 70, 155; and Ashton, Ruth, 59, 66, 68; politics of, 69; post-Documentary Unit career of, 66, 104, 135–36; and radio research, 61, 67, 69; and recordings, 64, 157; and *Red Channels*, 131, 135–36; and television, 66, 136, 156
Herman, George, 144
Hermann, Bernard, 25
Hersey, John, 28, 80, 102, 119
Hicks, George, 30, 80–82, 85–87
Hilmes, Michele, 4, 101, 157–58
Hindenburg disaster, 30, 96
Hiroshima (Hersey), 28, 80, 102, 119
Hiroshima (radio program), 28–29, 80, 102, 116

Hitler, Adolf, 16, 19, 38, 134; and *The March of Time* (radio series), 14; and McKee, Elmore, 108, 109; and *Quick and the Dead*, 120; and *They Fly through the Air with the Greatest of Ease*, 24
Ho Chi Minh, 113
Hollenbeck, Don: and *CBS Views the Press*, 48, 139; and *Hear It Now*, 139, 140, 141, 149; and *You Are There* (radio series), 94, 96, 103
Hollywood Fights Back, 44
Hollywood Quarterly, 39, 43
Home Is What You Make It, 9, 71–73, 112, 124
Hook, Sidney, 107, 108
Hoover, Herbert, 147
Hope, Bob, 3, 9, 56; and *One World Flight*, 40–41, 178n89; and *Quick and the Dead*, 117–27
Hottelet, Richard C., 33, 96, 103
House Committee on Un-American Activities (HUAC): and Corwin, Norman, 44, 130, 179n107; and Hollywood, 92, 99, 130, 134, 136–38; and radio commentators, 22; and Schlesinger, Arthur Jr., 130
House I Live In, The (radio play), 18
Howe, Quincy, 22, 94, 96, 98, 103
Hucksters, The (Wakeman), 47
Hughes, Langston, 15, 19, 201n18
Hume, Paul, 140
Hurst, Fannie, 29
Huston, John, 18
Hutchins Commission, 23
Hutchinson, Anne, 99

I Can Hear It Now, 7, 9, 117, 124, 126, 151–52; creation of, 114–16, 196n439
Ickes, Harold, 51
Independent Citizens Committee of the Arts, Sciences, and Professions, 69, 174n21
Industrial Areas Foundation (IAF), 52
Information Please, 54
International Ladies' Garment Workers' Union, 88

Jaffe, Sam, 135
"Japanese Americans," 18
Jefferson, John, 144
Johnson, Laurence, 133, 137
Johnstone, G. W. "Johnny," 16

Jolson, Al, 31
Jones, Hardin, 59
Julian, Joseph, 89, 131, 134–35
juvenile delinquency and documentary, 2, 48, 81, 144; and *The Eagle's Brood*, 3, 9, 51–56, 62–64, 68

Kaltenborn, H. V., 21
Kefauver hearings on organized crime, 10, 143–44
Keith, Michael, 4
Kesten, Paul, 46, 48
KIDO (Boise, Idaho), 21
Kielce pogrom, 33, 37
Kirkpatrick, Theodore, 131–32
Klauber, Ed, 16, 96
Kleinsorge, Father Wilhelm, 119, 121
Kobak, Edgar, 102
Korean War, 110, 133, 160; and *Hear It Now*, 10, 140, 144–47, 149, 153, 157–58; and *Living*, 113; and *Quick and the Dead*, 10, 119, 124, 126, 127; and *See It Now*, 151

labor relations and documentary, 15; and *Cavalcade of America*, 14; and *Communism—U.S. Brand*, 89; and *Hear It Now*, 144; and *Living*, 83; and *People Act*, 110, 112
LaMotta, Jake, 143
Lampell, Millard, 201n18
Lancaster, Burt, 134
Land is Bright, The, 50
Lange, Dorthea, 15
Larkin, John, 76
Last Day of the War, The, 18
Laurence, William, 115, 116–27
Laurents, Arthur, 15, 18, 201n18
Lauter, Bob, 90–91
Lawrence, Jerome, 39, 43
Lazarsfeld, Paul, 9, 61–64, 67, 69
Leader, Anton, 88, 89
Leahy, William, 149
Ledford, Paul, 30
Lee, Gypsy Rose, 132–33
Lee, Robert E., 95, 98
Légaré, Jean-Louis, 98
Lenthall, Bruce, 15, 157
Leonard, Bill, 140, 141
LeSueur, Larry, 167n34

Let Us Now Praise Famous Men (Agee and Evans), 15
Lewis, Bob, 121
Lewis, Fulton Jr., 21
liberalism: and Alinsky, Saul, 52; and anticommunism, 10, 101–2, 106–8, 124–25, 153, 158; and CBS, 10, 24–25, 45, 67–70, 131; and Corwin, Norman, 24–27, 69, 101–2, 131, 164; and "Cultural Front," 5, 6–7; and Murrow, Edward R., 21, 70, 138, 152–54; and One World, 27–29, 72, 102, 104–5, 153, 158; and prewar documentary, 14–16, 18–19; and radio commentators, 21–22, 41; and "reformist Left," 5, 161; and Shayon, Robert Lewis, 69–70, 98–99, 106–7, 131, 136–37, 161; and Wallace, Henry, 26–27, 91, 106; and *You Are There* (radio series), 98–99. See also *Communism—U.S. Brand*; *Counterattack: Red Channels*
Library of Congress, 4, 30
Life, 51
Lilienthal, David, 127–28; and *Quick and the Dead*, 119, 120, 122; and *Sunny Side of the Atom*, 58, 60
Lincoln, Abraham, 64, 93–95, 96, 98, 161
Lincoln, Mary Todd, 94, 95
Living, 79, 124, 131–32, 161, 194–95n21; and "Malice in Wonderland," 113–14, 132, 158. See also *Living 1948*; *People Act, The*
Living 1948, 9, 72–79, 97–98, 100–102, 156
Lomax, Alan, 4, 30, 201n18
Lomax, John, 4, 30
London Blitz, 1, 16, 70, 147
Long Life and a Merry One, A, 57, 64
Lorentz, Pare, 15
Luce, Henry, 26, 40
Lukas, Paul, 119, 120, 121
Lux Radio Theater, 59
Lydgate, William, 75

MacArthur, Douglas, 10, 34, 140, 146–47, 149
MacLeish, Archibald, 26; and *Fall of the City*, 16; and Murrow, Edward R., 1, 17; and *Operation Crossroads*, 51, 69, 162
"Malice in Wonderland," 113–14, 132, 148, 158
Manoff, Robert Karl, 127
Mantle, Mickey, 143
Marble, Harry, 96
March of Time, The (film newsreel), 14

March of Time, The (radio series), 3, 8, 13–14, 30
Marshall, George, 149–50, 153
Marshall Plan, 98, 149
Martin, Robert P., 144
McCarran, Pat, 106–7
McCarthy, Joseph, 6, 130, 148–49; and Murrow, Edward R., 10, 125, 138, 148–51, 153–54, 158, 161
McConnell, Joseph, 119, 125
McKee, Elmore, 10, 107–13, 125–27
Meitner, Lise, 119, 120
Memphis Belle (documentary film), 18
Meyer, Agnes, 52
Michel, Werner, 66
Mickelson, Sig, 123–24, 130, 135, 138, 139
Miller, Arthur, 15, 201n18
Miller, Justin, 48
Miller, Merle, 29
Mind in the Shadow, 66
minorities and documentary, 6, 18–19, 101, 159; and *Among Ourselves*, 65; and *Arrows in the Dust*, 185n95; and *Hear It Now*, 144, 145; and *One World Flight*, 27–28, 39; and *Open Letter on Race Hatred*, 18–19, 174n26; and *People Act*, 111; and *To Secure These Rights*, 101–2; and *You Are There* (radio series), 98
Mister Ledford and the TVA, 30
"Mobilization against Thought Police in the USA," 22, 179n107
Montgomery, Bernard, 27, 168n34
Moorehead, Agnes, 9, 14, 59, 61
Morrison, Herbert, 30
Moss, Annie Lee, 151
Muir, Jean, 135
Murrow, Edward R., 6–7, 127–28, 158; and *As Others See Us*, 75; and CBS Documentary Unit, 2, 45, 48–59, 66–70; CBS executive career of, 23, 32, 46–48, 65, 93; and *CBS Views the Press*, 48; and Corwin, Norman, 25, 42; death of, 6; and *Eagle's Brood*, 50–56; and editorializing, 70; Friendly initial pairings with, 10, 114–15, 123–24, 139; and *Hear It Now*, 3, 10, 139–58; and *I Can Hear It Now*, 7, 9, 114–15, 117, 124, 126, 150–51; liberalism of, 21, 70, 153; and Korean War, 10, 140, 144–47, 151, 156–57; and McCarthy, Joseph, 10, 125, 138, 148–50, 153–54, 161; and objectivity, 16–17, 138, 152–53, 156–57; and Paley, William, 46–47, 53, 139; and *Person to Person*, 142; and prostitution documentary, 152; and red-baiting, 125, 135, 136, 138; and *Report to the Nation—The 1950 Elections*, 139; and RTDNA speech, 151, 163; and *See It Now*, 2–3, 10, 140, 145, 150–56, 161; and Shayon, Robert Lewis, 1–2, 6, 50–56, 96, 138, 167–68n34; and *Sunday with Murrow*, 138; and *Sunny Side of the Atom*, 58–59, 68, 70; and television, 10, 21, 124, 151; and U.S. Information Agency, 148; and *Who Killed Michael Farmer?*, 152; and World War II, 1–2, 16–18, 25, 31, 96, 100; and *You Are There* (radio series), 9, 93, 96, 100, 114
Mussolini, Benito, 14, 16, 24, 38
Mutual Broadcasting System, 16, 131; and documentaries, 8, 79, 101–2; and radio commentators, 21; and recordings, 31, 80

Nation, The, 40, 70, 123
National Association of Broadcasters (NAB), 48, 159
National Opinion Research Center, 19–20
National Social Welfare Agency, 195n21
Nation's Nightmare, The, 151
NBC (National Broadcasting Company) 40, 87; and Blue Network, 80; news operation of, 16, 20; postwar public affairs series of, 20–21; and radio commentators, 21; and recording ban, 30–31, 53, 73, 105, 125; and red-baiting, 131–32, 135; and television, 21, 125. *See also* CBS–NBC rivalry; NBC documentary programs and series
NBC documentary programs and series, 8, 71, 83, 85, 142, 160; *Home Is What You Make It*, 9, 71–72, 124; *Living*, 79, 124, 131–32, 161, 194–95n21; *Living 1948*, 9, 72–79, 97–98, 100–102, 156; "Malice in Wonderland," 113–14, 132, 148, 158; *People Act*, 10, 107–13, 125–26, 144, 158; *Quick and the Dead*, 10, 114–28; *Voices and Events*, 105, 124; *Yesterday, Today, and Tomorrow*, 148–49
"Negro Domestic, The," 18, 21
Negro Soldier, The, 18
Nehru, Pandit, 33, 35, 38
Nerone, John, 125, 156

New Leader, 135
New Republic, 41, 91
NewsActing, 13
NewsCasting, 13
News of America, 20
News of the World, 20
Newsweek, 40, 61, 79
New World A-Coming, 18, 21
New Yorker, 28, 61–62, 87; Corwin, Norman and, 25; Murrow, Edward R. and, 65, 153
New York Herald Tribune, 52, 75, 93, 99, 117, 139, 141
New York Times, 108, 138; and *Hear It Now*, 143; and *Living 1948*, 75, 76; and 1948 presidential election, 78; and *Open Letter on Race Hatred*, 19; and *Red Channels*, 135; and *Sunny Side of the Atom*, 61. *See also* Gould, Jack; Laurence, William
New York World, 14
Nichols, Kenneth, 118
1960?? Jiminy Cricket!, 9, 82–84, 100, 157
Nord, David Paul, 5–6, 160–61
nuclear energy and weapons. *See* atomic energy and weapons in documentary
Nuremberg Trials, 48–49

Oak Ridge National Laboratory, 58, 59, 120, 122
objectivity: and editorializing, 70; and Murrow, Edward R., 16–17, 138, 152–53, 156–57; and radio news, 16, 20
Oboler, Arch, 15, 18
On a Note of Triumph, 8, 25–27, 28, 40, 43, 68
One World (Willkie), 8, 27
One World Flight, 3, 48, 104, 152, 157; discussion of, 28–45, 163–64; and recordings, 8–9, 29–39, 42, 53, 73; and red-baiting, 44–45, 106, 130; and scheduling controversy, 9, 40–42, 45, 56, 178n89
Open Letter on Race Hatred, 2, 18–19, 131, 174n26
Operation Crossroads, 62, 69, 116; airing of, 29, 50–51; Einstein, Albert and, 51, 58, 162
Operation Idaho, 21
Oppenheimer, Robert, 58, 60
O'Shea, Daniel, 133
Our Foreign Policy, 20–21
Our Miss Brooks, 104

Packard, Vance, 62
Paine, Thomas, 161
Paley, William, 51, 105, 124, 137; and Blue Book, 9, 41, 47–48, 159–60; and CBS Documentary Unit, 9, 45–48, 50, 66; and Corwin, Norman, 10, 28, 31, 34, 67, 104; and *Hear It Now*, 139, 150; and Murrow, Edward R., 46–47, 53, 139; and NBC rivalry, 16, 46, 67; and *You Are There* (radio series), 93
Parsons, William "Deak," 118, 121
Passport for Adams, 30
Pearson, Drew, 21, 148, 149
People Act, The, 10, 107–13, 125–26, 144, 151, 158
People Look at Radio, The (Columbia University Bureau of Applied Research), 69, 171n41
People, Yes, The (Sandburg), 140–41, 151
Perl, Arnold: and CBS, 49, 65, 66, 79; and Mutual, 79, 101–2; and *Red Channels*, 131
Person to Person, 142
Petrillo, James, 35
Pierpoint, Robert, 145, 146
Plow That Broke the Plains, The, 15
PM, 40, 56
Polk, George, 32, 176n47
Porter, Paul, 23
poverty and documentary, 72, 80–82, 104–5, 111
Prelude to War, 18
Priestley, J. B., 33
Program Analyzer. *See* Stanton, Frank: and Lazarsfeld-Stanton Program Analyzer
Progressive Citizens of America (PCA), 44, 106; and Shayon, Robert Lewis, 69–70, 106, 129, 136–37
Progressive party, 10, 91, 129
Prokofiev, Sergei, 33
Public Service Responsibility of Broadcast Licensees (Federal Communications Commission). *See* Blue Book
Pyle, Ernie, 26

Quick and the Dead, The, 3, 10, 114–28

race relations and documentary. *See* minorities and documentary
radio commentators, 21, 22, 41

Radio Drama in Action (Barnouw), 133
radio reform. *See* reform movements in radio
radio research, 61–64, 68–69
Radio's Second Chance (Siepmann), 47, 70
Radio-Television News Directors Association (RTDNA), 163
Radio Writers Guild, 136
Radulovich, Milo, 151, 153
RCA (Radio Corporation of America), 21, 24
Reader's Digest, 52
Ream, Joseph, 131
recordings: and ABC, 31, 80–82, 85, 100, 111; and audiotape, 10, 31, 105–6, 109, 114–15, 125; ban against, 30–31, 53, 73, 105, 125, 175n39; and documentary form, 5, 53, 155–59; early use in radio, 30–31; and *Hear It Now*, 139, 141–42, 152; and *I Can Hear It Now*, 114–15, 124; and Mutual, 31, 80; and *One World Flight*, 8–9, 29–39, 42, 53, 73; and *People Act*, 109–11; and *Quick and the Dead*, 117–19, 122, 124; and *Voices and Events*, 105, 124; and *We Went Back*, 64–65; and wire recorder, 8–9, 31–33, 53, 80–82
red-baiting. See *Counterattack*, House Committee on Un-American Activities (HUAC), *Red Channels*
Red Channels (American Business Consultants, Inc.), 2, 10, 130–37, 141
reform movements in radio, 7, 13, 22–23, 159–60, 162. *See also* Blue Book
Report Card, 66
Report to the Nation—The 1950 Elections, 139
Reveille for Radicals (Alinsky), 52
Rhymer, Paul, 173n16
Richman, T. Lefoy, 84
Ridgway, Matthew, 145
Roberts, Ken, 96, 98
Robeson, Paul, 28
Robinson, Hubbell, 66, 102–3
Robinson, Sugar Ray, 143, 150, 202n45
Robson, William, 2, 15, 18–19, 131
Rodman, Howard, 57
Roosevelt, Eleanor, 14
Roosevelt, Franklin D., 14, 21, 25–27, 120, 122
Rorty, Richard, 5, 161
Rose, Norman, 89
Rosenberg, Anna, 149, 153

Rosteck, Thomas, 153–54, 158
Rosten, Norman, 15, 18; and *Communism: U.S. Brand*, 9, 91–93, 102, 131; and *Red Channels*, 131
Russell, Harold, 22

Sandburg, Carl, 26, 64, 140–41, 151
Sarnoff, David, 118
Saturday Review, 136, 141
Saudek, Robert, 135; ABC career of, 9, 79–93, 100, 132; and Ford Foundation, 125, 160
Schechter, Abe, 16
Schlesinger, Arthur Jr., 10, 106–8, 130, 132
Schoenbrun, David, 143
School Teacher—1947, 80
Schudson, Michael, 152, 156, 161
Schwartau, Bill, 119
Schwartz, Tony, 4
Scott, Ed, 142, 143
See It Now, 2, 3, 140, 142, 145; debut of, 151; discussion of, 152–56, 158, 161
Seldes, Gilbert, 26
Selznick, David O., 53–54
Semmler, Alexander, 35
Sevareid, Eric, 1
Seward, William, 94
Shaw, G. Howland, 51–53
Shayon, Robert Lewis, 7, 11, 46, 67, 154; departure from CBS, 10, 105–6, 129; and *Eagle's Brood*, 3, 9, 50–57, 68–69, 107, 141, 156; early career of, 50–51; and *Hear It Now*, 141; liberalism of, 69–70, 98–99, 106–7, 131, 136–37, 161; and Murrow, Edward R., 1–2, 6, 50–56, 96, 138, 167–68n34; and *Operation Crossroads*, 50–51, 162; and radio research, 61–64; and red-baiting, 129–31, 136–38, 141; and *Saturday Review*, 136, 141; and *You Are There* (radio series), 3, 9, 65–66, 93–103, 114
Shirer, William, 1, 21, 65
Siepmann, Charles, 47, 48, 70, 114, 160
Sitting Bull, 98
Slums, 9, 80–82, 85, 100, 101, 111
Smith, Howard K., 1, 32
Smith, Red, 117, 119, 127
Socrates, 99
Sorry, Wrong Number (radio program), 173n16

South Pacific (stage musical), 118
Sperber, A. M., 153, 167n34
Squier, Susan Merrill, 4
Stalin, Joseph, 14, 27, 113, 134, 149; and *One World Flight*, 33, 42
Stanton, Edwin, 95
Stanton, Frank, 46, 47, 65, 124, 139, 150; and *Eagle's Brood*, 54, 56; and Lazarsfeld-Stanton Program Analyzer, 9, 61–64, 67
Stassen, Harold, 51
Steinbeck, John, 15
Sterling, Christopher, 4
Stimson, Henry L., 150
St. John, Robert, 21
Stott, William, 8, 15
Sunday with Murrow, 138
Sunny Side of the Atom, The, 3, 9, 68–70, 127, 162; origins and production of, 58–61; and radio research, 64
Superman (radio series), 25
sustaining programming, 8, 16, 32; and ABC documentaries, 80, 91; and Blue Book, 47–48, 67; and CBS Documentary Unit, 46–49; and *Hear It Now*, 139, 202–3n47; and *Living 1948*, 73; and *One World Flight*, 40–42, 44; and *You Are There* (radio series), 99–100
Swing, Raymond, 21

Taft, Charles, 109, 111–12
Taft, Robert, 109, 147, 187n8
tape. *See* recordings
Taylor, Davidson, 32, 93, 104–5
Taylor, Ruth Ashton. *See* Ashton, Ruth
television: and documentary, 3, 66, 150–51, 156, 158–60; and early network news operations, 21; growth of, 2, 105, 190, 125; and Kefauver hearings, 143–44. See also *See It Now*
They Call Me Joe, 18
They Came This Way, 20
They Fly through the Air with the Greatest of Ease, 24–25
This is War!, 25, 50
Thomas, J. Parnell, 92
Thomas, Lowell, 21
Thomas, Norman, 187n8
Thompson, "Big Bill," 14
"Thought Control in the USA," 44
Tibbets, Paul, 118, 121

Tifft, Susan, 152
Time, 16, 26; and *Communism—U.S. Brand*, 90; and Corwin, Norman, 25, 40; and *Eagle's Brood*, 56; and *March of Time* (radio series), 8, 13–14; and *Open Letter on Race Hatred*, 19; and *Quick and the Dead*, 118, 119; and *You Are There* (radio series), 101
Tobey, Charles, 29
Tolstoy, Leo, 95
To Secure These Rights, 101–2
Toussaint L'ouverture, 95, 98
Trammell, Niles, 21
Trout, Robert, 57, 115, 119, 121
Truman, Harry S.: and atomic weapons, 115, 116, 121; and civil rights, 101–2; and *Hear It Now*, 140, 146–47, 149; and juvenile delinquency, 51; 1948 election victory of, 78–79; and *One World Flight*, 32, 36
Twentieth Century Fund, 10, 83, 108–12
Twenty-six by Corwin, 24

Unhappy Birthday, 29, 58, 80, 116
United Nations, 44, 76, 104, 140, 149
United States Atomic Energy Commission, 118
Urey, Harold, 51, 121, 126–27
U.S. Information Agency, 148
U.S. News and World Report, 126

Vandenberg, Arthur, 147
Vandercook, John, 21
Variety, 11, 32, 84, 104, 105; and anticommunism in broadcasting, 113, 130; and *Atom and You*, 79; and Blue Book, 47–48; and CBS Documentary Unit, 49, 65; and *Communism: U.S. Brand*, 87, 90; and Corwin, Norman, 26, 35, 39–42, 44; and *Eagle's Brood*, 82; and Friendly, Fred, 150; and Grauer, Ben, 132; and *Hear It Now*, 139; and Heller, Robert, 136; and House Committee on Un-American Activities (HUAC), 22, 44; and *Living 1948*, 73, 74, 100; and *Long Life and a Merry One*, 57; and *1960?? Jiminy Cricket!*, 84; and *People Act*, 112–13; and *Quick and the Dead*, 119, 122–23; and radio commentators, 21, 22; and recordings, 65, 156; and Shayon, Robert Lewis, 50; and *Slums*, 82; and *Sunny Side of the Atom*, 61; and television, 105,

150; and *We Went Back*, 65; and *You Are There* (radio series), 93, 96, 99
V.D.: The Conspiracy of Silence, 3, 9, 84–87, 100, 101
Vic and Sade, 173n16
Vital Center, The (Schlesinger), 10, 106, 107
Voice of America, 142, 148
Voices and Events, 105, 124

Walker, Paul, 68
Wallace, Henry, 51; and Corwin, Norman, 26, 106, 174n21; and *New Republic*, 41, 91; political philosophy of, 26–27, 43, 106; presidential candidacy of, 10, 44, 79, 129, 187n8
War of the Worlds, The (radio program), 94, 96–97
Warren, Earl, 187n8
Warsaw ghetto, 18, 33, 37, 87, 91–92
Washington Post, 52
WEAN (Providence, Rhode Island), 114
Welles, Gideon, 94–95
Welles, Orson, 14, 15, 94, 201n18
Wershba, Joe, 142, 143
We Went Back, 64–65
What's My Line?, 96
White, Llewellyn, 67
White, Paul, 16, 17, 57–58
Whitman, Walt, 26
Who Killed Michael Farmer?, 152
Who Said That?, 115, 123, 199n79
Why We Fight, 18, 50
Willkie, Wendell, 8, 27–28, 43, 174n26

Winchell, Walter, 21, 22
wire recorder. *See* recordings
Wishengrad, Morton, 10, 18, 136; and *Communism—U.S. Brand*, 9, 87–93, 102, 106; and *Red Channels*, 131–32, 135
WMCA (New York), 18, 21
Woman's Home Companion, 52
women and documentary, 6, 101, 159; and Ashton, Ruth, 57–58, 68; and Dick, Elsie, 79, 188n28; and *Hear It Now*, 143; and *Living 1948*, 77–78; and *People Act*, 112; and *You Are There* (radio series), 99
Works Progress Administration, 15
World News Roundup, 20
World War II: and CBS, 1–2, 16–19; and Corwin, Norman, 24–27, 30; and documentary and news programming, 6–7, 16–19, 28–29, 30; and Murrow, Edward R., 1–2, 16–18, 25, 31, 96, 100
Wyler, William, 18

Yesterday, Today and Tomorrow, 148
You Are There (radio series), 3, 5, 9, 65–66, 157; cancellation of, 105, 193n4; criticism of, 9, 93, 95, 114; origins and production of, 93–101, 191n78; parodies of, 95, 103
You Are There (TV series), 192n102, 193n4
Your Children Today, 79
Young, Robert, 30
Young and Rubicam, 40

Zousmer, Jesse, 142

THE HISTORY OF COMMUNICATION

Selling Free Enterprise: The Business Assault on Labor and Liberalism, 1945–60
   Elizabeth A. Fones-Wolf
Last Rights: Revisiting Four Theories of the Press   *Edited by John C. Nerone*
"We Called Each Other Comrade": Charles H. Kerr & Company, Radical Publishers
   Allen Ruff
WCFL, Chicago's Voice of Labor, 1926–78   *Nathan Godfried*
Taking the Risk Out of Democracy: Corporate Propaganda versus Freedom and
   Liberty   *Alex Carey; edited by Andrew Lohrey*
Media, Market, and Democracy in China: Between the Party Line and the Bottom Line
   Yuezhi Zhao
Print Culture in a Diverse America   *Edited by James P. Danky and Wayne A. Wiegand*
The Newspaper Indian: Native American Identity in the Press, 1820–90
   John M. Coward
E. W. Scripps and the Business of Newspapers   *Gerald J. Baldasty*
Picturing the Past: Media, History, and Photography   *Edited by Bonnie Brennen and
   Hanno Hardt*
Rich Media, Poor Democracy: Communication Politics in Dubious Times
   Robert W. McChesney
Silencing the Opposition: Antinuclear Movements and the Media in the Cold War
   Andrew Rojecki
Citizen Critics: Literary Public Spheres   *Rosa A. Eberly*
Communities of Journalism: A History of American Newspapers and Their Readers
   David Paul Nord
From Yahweh to Yahoo!: The Religious Roots of the Secular Press   *Doug Underwood*
The Struggle for Control of Global Communication: The Formative Century   *Jill Hills*
Fanatics and Fire-eaters: Newspapers and the Coming of the Civil War
   Lorman A. Ratner and Dwight L. Teeter Jr.
Media Power in Central America   *Rick Rockwell and Noreene Janus*
The Consumer Trap: Big Business Marketing in American Life   *Michael Dawson*
How Free Can the Press Be?   *Randall P. Bezanson*
Cultural Politics and the Mass Media: Alaska Native Voices   *Patrick J. Daley and
   Beverly A. James*
Journalism in the Movies   *Matthew C. Ehrlich*
Democracy, Inc.: The Press and Law in the Corporate Rationalization of the Public
   Sphere   *David S. Allen*
Investigated Reporting: Muckrakers, Regulators, and the Struggle over Television
   Documentary   *Chad Raphael*
Women Making News: Gender and the Women's Periodical Press in Britain
   Michelle Tusan
Advertising on Trial: Consumer Activism and Corporate Public Relations in the 1930s
   Inger Stole

Speech Rights in America: The First Amendment, Democracy, and the Media
  Laura Stein
Freedom from Advertising: E. W. Scripps's Chicago Experiment   *Duane C. S. Stoltzfus*
Waves of Opposition: The Struggle for Democratic Radio, 1933–58
  Elizabeth Fones-Wolf
Prologue to a Farce: Democracy and Communication in America   *Mark Lloyd*
Outside the Box: Corporate Media, Globalization, and the UPS Strike   *Deepa Kumar*
The Scripps Newspapers Go to War, 1914–1918   *Dale Zacher*
Telecommunications and Empire   *Jill Hills*
Everything Was Better in America: Print Culture in the Great Depression
  David Welky
Normative Theories of the Media   *Clifford G. Christians, Theodore Glasser,
  Denis McQuail, Kaarle Nordenstreng, Robert A. White*
Radio's Hidden Voice: The Origins of Public Broadcasting in the United States
  Hugh Richard Slotten
Muting Israeli Democracy: How Media and Cultural Policy Undermine Free
  Expression   *Amit M. Schejter*
Key Concepts in Critical Cultural Studies   Edited by *Linda Steiner and
  Clifford Christians*
Refiguring Mass Communication: A History   *Peter Simonson*
Radio Utopia: Postwar Audio Documentary in the Public Interest   *Matthew C. Ehrlich*

MATTHEW C. EHRLICH is a professor of journalism at the University of Illinois at Urbana-Champaign and the author of *Journalism in the Movies*.

The University of Illinois Press
is a founding member of the
Association of American University Presses.

Composed in 10/13.5 Adobe Minion Pro
with FF Meta display
at the University of Illinois Press
Manufactured by Sheridan Books, Inc.

University of Illinois Press
1325 South Oak Street
Champaign, IL 61820-6903
www.press.uillinois.edu